156
Advances in Polymer Science

Editorial Board:
A. Abe · A.-C. Albertsson · H.-J. Cantow
K. Dušek S. Edwards · H. Höcker
J. F. Joanny · H.-H. Kausch · K.-S. Lee
J. E. McGrath · L. Monnerie · S. I. Stupp
U. W. Suter · G. Wegner · R. J. Young

Springer
Berlin
Heidelberg
New York
Barcelona
Hong Kong
London
Milan
Paris
Tokyo

Molecular Simulation
Fracture
Gel Theory

With contributions by
H. R. Brown, C. Creton, C.-Y. Hui, W. H. Jo,
E. J. Kramer, K. Suematsu, J. S. Yang

 Springer

This series presents critical reviews of the present and future trends in polymer and biopolymer science including chemistry, physical chemistry, physics and materials science. It is addressed to all scientists at universities and in industry who wish to keep abreast of advances in the topics covered.

As a rule, contributions are specially commissioned. The editors and publishers will, however, always be pleased to receive suggestions and supplementary information. Papers are accepted for „Advances in Polymer Science" in English.

In references Advances in Polymer Science is abbreviated Adv Polym Sci and is cited as a journal.

Springer APS home page: http://link.springer.de/series/aps/ or
http://link.springer-ny.com/series/aps/
Springer-Verlag home page: http://www.springer.de

ISSN 0065-3195
ISBN 3-540-42126-2
Springer-Verlag Berlin Heidelberg New York

Library of Congress Catalog Card Number 61642

This work is subject to copyright. All rights are reserved, whether the whole or part of the material is concerned, specifically the rights of translation, reprinting, re-use of illustrations, recitation, broadcasting, reproduction on microfilms or in other ways, and storage in data banks. Duplication of this publication or parts thereof is only permitted under the provisions of the German Copyright Law of September 9, 1965, in its current version, and permission for use must always be obtained from Springer-Verlag. Violations are liable for prosecution under the German Copyright Law.

Springer-Verlag Berlin Heidelberg New York
a member of BertelsmannSpringer Science+Business Media GmbH

© Springer-Verlag Berlin Heidelberg 2002
Printed in Germany

http://www.springer.de

The use of registered names, trademarks, etc. in this publication does not imply, even in the absence of a specific statement, that such names are exempt from the relevant protective laws and regulations and therefore free for general use.

Typesetting: Data conversion by MEDIO, Berlin
Cover: MEDIO, Berlin
Printed on acid-free paper   SPIN: 10753281   02/3020hu - 5 4 3 2 1 0

Editorial Board

Prof. Akihiro Abe
Department of Industrial Chemistry
Tokyo Institute of Polytechnics
1583 Iiyama, Atsugi-shi 243-02, Japan
E-mail: aabe@chem.t-kougei.ac.jp

Prof. Ann-Christine Albertsson
Department of Polymer Technology
The Royal Institute of Technolgy
S-10044 Stockholm, Sweden
E-mail: aila@polymer.kth.se

Prof. Hans-Joachim Cantow
Freiburger Materialforschungszentrum
Stefan Meier-Str. 21
D-79104 Freiburg i. Br., FRG
E-mail: cantow@fmf.uni-freiburg.de

Prof. Karel Dušek
Institute of Macromolecular Chemistry, Czech
Academy of Sciences of the Czech Republic
Heyrovský Sq. 2
16206 Prague 6, Czech Republic
E-mail: dusek@imc.cas.cz

Prof. Sam Edwards
Department of Physics
Cavendish Laboratory
University of Cambridge
Madingley Road
Cambridge CB3 OHE, UK
E-mail: sfe11@phy.cam.ac.uk

Prof. Hartwig Höcker
Lehrstuhl für Textilchemie
und Makromolekulare Chemie
RWTH Aachen
Veltmanplatz 8
D-52062 Aachen, FRG
E-mail: hoecker@dwi.rwth-aachen.de

Prof. Jean-François Joanny
Institute Charles Sadron
6, rue Boussingault
F-67083 Strasbourg Cedex, France
E-mail: joanny@europe.u-strasbg.fr

Prof. Hans-Henning Kausch
c/o IGC I, Lab. of Polyelectrolytes
and Biomacromolecules
EPFL-Ecublens
CH-1015 Lausanne, Switzerland
E-mail: kausch.cully@bluewin.ch

Prof. Kwang-Sup Lee
Department of Macromolecular Science
Hannam University
Teajon 300-791, Korea
E-mail: kslee@eve.hannam.ac.kr

Prof. James E. McGrath
Polymer Materials and Interfaces Laboratories
Virginia Polytechnic and State University
2111 Hahn Hall
Blacksbourg
Virginia 24061-0344, USA
E-mail: jmcgrath@chemserver.chem.vt.edu

Prof. Lucien Monnerie
École Supérieure de Physique et de Chimie
Industrielles
Laboratoire de Physico-Chimie
Structurale et Macromoléculaire
10, rue Vauquelin
75231 Paris Cedex 05, France
E-mail: lucien.monnerie@espci.fr

Prof. Samuel I. Stupp
Department of Measurement Materials Science
and Engineering
Northwestern University
2225 North Campus Drive
Evanston, IL 60208-3113, USA
E-mail: s-stupp@nwu.edu

Prof. Ulrich W. Suter
Department of Materials
Institute of Polymers
ETZ,CNB E92
CH-8092 Zürich, Switzerland
E-mail: suter@ifp.mat.ethz.ch

Prof. Gerhard Wegner
Max-Planck-Institut für Polymerforschung
Ackermannweg 10
Postfach 3148
D-55128 Mainz, FRG
E-mail: wegner@mpip-mainz.mpg.de

Prof. Robert J. Young
Manchester Materials Science Centre
University of Manchester and UMIST
Grosvenor Street
Manchester M1 7HS, UK
E-mail: robert.young@umist.ac.uk

Advances in Polymer Science
Now Also Available Electronically

For all customers with a standing order for Advances in Polymer Science we offer the electronic form via LINK free of charge. Please contact your librarian who can receive a password for free access to the full articles. By registration at:

http://link.springer.de/series/aps/reg_form.htm

If you do not have a standing order you can nevertheless browse through the table of contents of the volumes and the abstracts of each article at:

http://link.springer.de/series/aps/
http://link.springer-ny.com/series/aps/

There you will find also information about the

- Editorial Bord
- Aims and Scope
- Instructions for Authors

Contents

Molecular Simulation Approaches for Multiphase Polymer Systems
W. H. Jo, J. S. Yang 1

Adhesion and Fracture of Interfaces Between Immiscible Polymers: From the Molecular to the Continuum Scale
C. Creton, E. J. Kramer, H. R. Brown, C.-Y. Hui 53

Recent Progress of Gel Theory: Ring, Excluded Volume, and Dimension
K. Suematsu 137

Author Index Volumes 101–156 215

Subject Index 227

Molecular Simulation Approaches for Multiphase Polymer Systems

Won Ho Jo, Jae Shick Yang

Hyperstructured Organic Materials Research Center and Department of Fiber and Polymer Science, Seoul National University, Seoul 151-742, Korea
e-mail: whjpoly@plaza.snu.ac.kr

Computer simulation has recently attracted much attention from both academia and industry because it provides much useful information on polymers, particularly multiphase polymer systems, which are not easily obtained by experiment. The computer simulation approaches for multiphase polymer systems may be classified into three modeling methods according to their levels of approximation, i.e., atomistic, coarse-grained, and atomistic-continuum modeling. In this article the three methods are applied to miscibility of polymer blends, compatibilizing effect of block copolymers, and mechanical properties of semicrystalline polymers, respectively, and their results are compared with experimental ones.

Keywords. Molecular simulation, Monte Carlo simulation, Block copolymer, Semicrystalline polymer

1	Introduction	2
2	**Miscibility of Polymer Blends**	3
2.1	Theoretical Background	4
2.1.1	Copolymer Blend Theory	4
2.1.2	Lattice Fluid Theory	5
2.2	Model and Simulation	7
2.3	Results and Discussion	9
2.3.1	Homopolymer/Copolymer Blends	9
2.3.2	Homopolymer/Homopolymer Blends	12
3	**Block Copolymer as a Compatibilizer**	18
3.1	Simulation Methods	19
3.2	Results and Discussion	21
3.2.1	A/B Binary Blends	21
3.2.2	A/B/C-b-D Ternary Blends	26
3.2.2.1	Effects of the Interaction Energy of Block Copolymers	26
3.2.2.2	Effects of the Chain Length of Block Copolymers	29
3.2.2.3	Effects of the Composition of Block Copolymers	33
3.2.2.4	Scaling	36

4	Mechanical Properties of Semicrystalline Polymers	40
4.1	Description of the Atomistic-Continuum Model	41
4.2	Simulation Methods	43
4.3	Results and Discussion	44
5	Conclusions	46
References		49

1
Introduction

In recent years, molecular simulation methods [1–5] including molecular mechanics (MM), molecular dynamics (MD), and Monte Carlo (MC) simulation have been applied to polymer systems. Molecular simulation provides a bridge between models and experiments, as a method using mathematical models to perform an analysis by means of computers. Therefore, molecular simulation has become a powerful tool in polymer science, complementing both theory and experiment. On the other hand, molecular simulation still has limitations in both space and time scales, which become more serious in dealing with multiphase polymer systems. The field of multiphase polymer materials includes all cases where the system consists of at least two phases, such as immiscible polymer blends, semicrystalline polymers, etc. Thus multiphase polymers have much larger scale in space and longer scale in time than full atomistic modeling can directly deal with.

Some alternatives have been developed to overcome these problems. For example, the coarse-grained lattice model [6] has been successful in extending the scope. In this model, polymer chains are represented by random walks on a lattice, and thus many states can be easily generated and equilibrated. Therefore, the lattice model has been widely used from polymer solutions and dense mixtures. However, this model does not include detailed information on the structure of materials and thus will be fictitious. Recently, multiscale simulation methods have been developed, which combines the atomistic and continuum levels. Tadmor et al. [7, 8] developed a quasicontinuum method to model the process involving multiple length scale such as nanoindentation. In this method, interatomic interactions are incorporated in the model via the crystal calculation based on the local state of deformation. Shenoy et al. [9] reformulated this model to examine the interactions between grain boundaries, dislocations, and cracks. Rafii-Tabar et al. [10] proposed a multi-scale seamless model of brittle crack propagation which couples the crack at the macroscales and nanoscales through an intermediate mesoscale continuum.

In this article, three modeling methods, classified by their levels of approximation, i.e., atomistic, coarse-grained, and atomistic-continuum modeling, are applied to miscibility of polymer blends, compatibilizing effect of block copolymers, and mechanical properties of semicrystalline polymers, respectively. In Sect. 2, the effect of the copolymer sequence distribution on the miscibility of homopolymer/copolymer blends is analyzed and the phase diagram of homopolymer/homopolymer blends as well as surface tension of individual polymers are predicted using an atomistic modeling technique. The phase separation kinetics of immiscible polymer blends containing various types of block copolymers are investigated via MC simulations based on a coarse-grained modeling in Sect. 3. Section 4 briefly describes the atomistic-continuum model proposed by Santos et al. [11] and its application to mechanical properties of a semicrystalline polymer. Conclusions are drawn in Sect. 5.

2
Miscibility of Polymer Blends

It is possible to obtain polymer blends with more desirable properties by mixing miscible polymers, and thus it is very important to examine the factors affecting the miscibility of polymer mixtures. The miscibility of homopolymer/copolymer blends has been successfully described by the binary interaction model [12–14]. One of successful applications of this model is to investigate the effect of the composition of the copolymer on the miscibility of poly(ethylene oxide) (PEO)/poly(styrene-co-acrylic acid) (SAA) blends [15] where both the specific interaction between ethylene oxide (EO) and acrylic acid (AA) segments and the intramolecular repulsive force in the SAA copolymer are responsible for the miscibility. The sequence distribution of the copolymer in homopolymer/copolymer blends may affect the charge distribution and the probability of contact between interaction sites, and consequently affect the miscibility of the blend. However, the binary interaction model is inadequate to study the sequence effect due to the assumption of a random distribution. Balazs et al. [16] showed that there is an optimal range of sequence distributions for which the homopolymer/copolymer system is miscible by using their own model. Therefore, it is necessary to re-examine the miscibility of PEO/SAA blend by taking into account the segment distribution of the SAA copolymer.

The lattice fluid equation-of-state theory for polymers, polymer solutions, and polymer mixtures is a useful tool which can provide information on equation-of-state properties, and also allows prediction of surface tension of polymers, phase stability of polymer blends, etc. [17–20]. The theory uses empty lattice sites to account for free volume, and therefore one may treat volume changes upon mixing, which are not possible in the Flory-Huggins theory. As a result, lower critical solution temperature (LCST) behaviors can, in principle, be described in polymer systems which interact chiefly through dispersion forces [17]. The equation-of-state theory involves characteristic parameters, p^*, v^*, and T^*, which have to be determined from experimental data. The least-squares fitting of density data as a function of temperature and pressure yields a set of parameters which best represent the data over the temperature and pressure ranges considered [21]. The method, however, requires tedious experiments to determine the characteristic parameters when dealing with a new polymer. Moreover, it may sometimes lead to a significant error in determining the characteristic parameters unless rigorous experimental data can be obtained. Thus, it would be helpful if the molecular simulation method could provide the characteristic parameters without the experimental efforts.

In this section, first, the effect of the copolymer sequence distribution in PEO/SAA blends is investigated by calculating the interaction energy parameters using atomistic modeling [22] and, second, the phase diagram of polystyrene (PS)/poly(vinyl methyl ether) (PVME) blends as well as the surface tensions of PS and PVME are predicted by the lattice fluid equation-of-state theory whose characteristic parameters are obtained using atomistic modeling [23].

2.1
Theoretical Background

2.1.1
Copolymer Blend Theory

For simplicity, the monomeric units of AA, styrene and EO are denoted as a, b, and c instead of using their full names. The free energy of mixing for a binary mixture of a homopolymer and a copolymer, $c_{N_1}/(a_f b_{1-f})_{N_2}$, is given by [12–14]

$$\frac{\Delta G}{RT} = \frac{\phi_1}{N_1}\ln\phi_1 + \frac{\phi_2}{N_2}\ln\phi_2 + \phi_1\phi_2\chi_{tot} \tag{1}$$

where χ_{tot} is the parameter that represents the strength of the polymer-polymer interaction. The homopolymer c_{N_1} has a volume fraction ϕ_1 and degree of polymerization N_1, and the copolymer $(a_f b_{1-f})_{N_2}$ has a volume fraction ϕ_2 and degree of polymerization N_2 with the composition of f and $1-f$. When the binary interaction model is invoked, the total interaction energy parameter χ_{tot} is given by

$$\chi_{tot} = f\chi_{ac} + (1-f)\chi_{bc} - f(1-f)\chi_{ab} \tag{2}$$

However, the simple binary interaction model is inadequate to study the sequence effects owing to its assumption of a random distribution. Assuming that the interaction energy parameters of all a-a and b-b pairs are equivalent and equal to zero and that all a-b interactions are equivalent to the average interaction parameter $\bar{\chi}_{ab}$, Balazs et al. [16] expressed χ_{tot} as the sum of the contribution of the composition χ_{comp} and the sequence distribution χ_{dist} as follows:

$$\chi_{tot} = \chi_{comp} + \chi_{dist} \tag{3}$$

$$\chi_{comp} = f_a\bar{\chi}_{ac} + f_b\bar{\chi}_{bc} - f_a f_b\bar{\chi}_{ab} \tag{4}$$

$$\chi_{dist} = \frac{f_{aa}^2}{f_a}\Delta\chi_a + \frac{f_{bb}^2}{f_b}\Delta\chi_b \tag{5}$$

where f_a and f_b are the fraction of a and b molecules in the copolymer, f_{aa}, f_{ab}, and f_{bb} are the pair probabilities of aa, ab, and bb pairs in a single chain, and thus $f_a = f_{aa} + f_{ab}$, $f_b = f_{bb} + f_{ba}$, and $f_{ab} = f_{ba}$. The $\overline{\chi}_{ac}$ is the average of the interactions between a occupying the central site in the triads and c, i.e., $\chi_{aaa:c}$, $\chi_{aab:c}$, and $\chi_{bab:c}$. The $\overline{\chi}_{bc}$ is similarly the average of $\chi_{aba:c}$, $\chi_{abb:c}$, and $\chi_{bbb:c}$. Although there are nine possible a-b interactions, all interactions will be assumed equivalent to the average interaction parameter $\overline{\chi}_{ab}$ which is the intramolecular interaction in the copolymer chain. Then, $\Delta\chi_a$ and $\Delta\chi_b$ are defined by Eqs. (6) and (7), respectively:

$$\Delta\chi_a = \chi_{aaa:c} - \overline{\chi}_{ac} \tag{6}$$

$$\Delta\chi_b = \chi_{bbb:c} - \overline{\chi}_{bc} \tag{7}$$

A negative $\Delta\chi_a$ implies that aaa-c interactions are energetically more favorable than any other type of a-c interactions, and a similar comment is applicable to $\Delta\chi_a$. The parameters θ and δ are introduced to specify the sequence distribution and the composition, respectively:

$$\theta = \frac{f_{ab}}{2f_a f_b} \tag{8}$$

$$f_a = \frac{1}{2}(1+\delta), f_b = \frac{1}{2}(1-\delta), \quad -1 \leq \delta \leq 1 \tag{9}$$

where θ can range from 0 for a block copolymer to 1 for an alternating copolymer, with the value 0.5 for a random copolymer. For symmetric copolymers, $\delta = 0$, and the copolymer composition can be expressed by δ.

2.1.2
Lattice Fluid Theory

For a polymer liquid, the lattice fluid equation-of-state can be written in a reduced form as

$$\tilde{\rho}^2 + \tilde{p} + \tilde{T}\left[\ln(1-\tilde{\rho}) + \left(1-\frac{1}{r}\right)\tilde{\rho}\right] = 0 \tag{10}$$

Assuming high molecular weight ($r \to \infty$) and near atmospheric pressure ($\tilde{p} \to 0$), the chemical potential μ is given by

$$\mu = -\tilde{\rho} + \tilde{T}\frac{(1-\tilde{\rho})\ln(1-\tilde{\rho})}{\tilde{\rho}} \tag{11}$$

where ϱ is density, p is pressure, v is volume, T is temperature, and r is molecular size parameter. The chemical potential, temperature, density, and pressure are reduced by their respective equation-of-state parameters as follows:

$$\tilde{\mu} = \frac{\mu}{rNkT^*}, \quad \tilde{T} = \frac{T}{T^*},$$
$$\tilde{\varrho} = \frac{1}{\tilde{v}} = \frac{v^*}{v}, \quad \tilde{p} = \frac{p}{p^*} \tag{12}$$

For polymer liquids, the gradient approximation in conjunction with the lattice fluid model has been used to calculate surface tensions [24, 25]. The Cahn-Hilliard relation for surface tension σ, in terms of reduced variables, can be expressed as

$$\tilde{\sigma} = 2\int_{\tilde{\varrho}_g}^{\tilde{\varrho}_l} (\tilde{\kappa}/\Delta\tilde{a})^{1/2} d\tilde{\varrho} \tag{13}$$

where

$$\Delta\tilde{a} = \tilde{\varrho}\left[\tilde{\mu}(\tilde{\varrho},\tilde{T}) - \tilde{\mu}(\tilde{\varrho}_l,\tilde{T})\right] \tag{14}$$

$$\sigma = \frac{\tilde{\sigma}}{(kT^*)^{1/3}(p^*)^{2/3}} \tag{15}$$

A dimensionless constant $\tilde{\kappa}$ has a priori theoretical value of 0.5 and the subscripts, g and l, in Eq. (13) denote the gas state and the liquid state, respectively. By substituting Eq. (11) into Eq. (14) and then integrating Eq. (13), the surface tension can be obtained. Such a procedure has been previously shown to yield estimates of surface tension for polyethylene melts in good agreement with experiment [26].

For simplicity, assuming that the close-packed volume of a PS mer (v_{PS}^*) is equal to that of a PVME mer (v_{PVME}^*), the binary polymer blend is miscible [20] when

$$f_{\phi\phi} = \left(\frac{\partial^2 f}{\partial \phi_{PS}^2}\right) = f_{\phi\phi,1} + f_{\phi\phi,2} < 0$$
$$f_{\phi\phi,1} = 2\tilde{\varrho}\phi_{PS}\phi_{PVME}\chi \tag{16}$$
$$f_{\phi\phi,2} = \tilde{\varrho}\phi_{PS}\phi_{PVME}\tilde{T}\Psi^2 p^*\beta$$

where f is the free energy per mer, ϕ_i is the volume fraction of component i, and β is the isothermal compressibility of the mixture. The interaction energy parameter χ is expressed as

$$\chi = \frac{\varepsilon^*_{PS} + \varepsilon^*_{PVME} - 2\varepsilon^*_{PS:PVME}}{kT} \tag{17}$$

where

$$\varepsilon^*_{PS} = kT^*_{PS}$$
$$\varepsilon^*_{PVME} = kT^*_{PVME} \tag{18}$$
$$\varepsilon^*_{PS:PVME} = \zeta\sqrt{\varepsilon^*_{PS}\varepsilon^*_{PVME}}$$

More detailed description for Eq. (16), including the definition of Ψ, can be found elsewhere [17, 20].

2.2
Model and Simulation

Segmental interaction energy parameters in PEO/SAA blends are calculated by docking methods [27, 28]. The interaction parameter $\chi_{i:j}$ can be calculated from a knowledge of pairwise interactions, w_{kl}s ($k, l \in \{i, j\}$), and coordination numbers, z_{kl}s by using Eq. (19):

$$\chi_{i:j} = \frac{1}{RT}\left\{\frac{1}{2}(z_{ij}w_{ij} + z_{ji}w_{ji}) - \frac{1}{2}(z_{ii} + w_{ii} + z_{jj}w_{jj})\right\} \tag{19}$$

where w_{kl} is an averaged pairwise interaction energy when a segment l is in contact with the center segment k. A similar explanation can be given for z_{kl}.

In order to simulate four w_{kl}s and z_{kl}s and to obtain $\chi_{i:j}$, the model segments i and j with proper charge distributions must be prepared. Nine model segments were prepared in this system: *aaa, aab, bab, aba, abb, bbb, a, b,* and *c*. For the triads, the partial charges were calculated utilizing the charge equilibration method [29] after constructing and optimizing the triad structure, and then the monomeric units on both sides were removed. Thus, a monomer structure, which has the charge distribution of a monomer in the center of the triad, was prepared. For example, the structure of a triad *aba* was produced, the partial charges were calculated, and then *a*s on both sides were removed, which produced a *b* segment with the charge distribution of the *b* monomer in the triad *aba*. The *b* segment will be denoted as *aba* in order to distinguish from other *b* segments which come from triads *abb* and *bbb*. The other model segments *aaa, aab, bab, abb,* and *bbb* were generated by following similar procedures. Three segments of *a* and *b* were used for calculating $\overline{\chi}_{ac}$ and $\overline{\chi}_{bc}$. Additionally, the model segments -*a*-, -*b*-, and *c* were made to have the total charge of zero, where the model segments -*a*- and -*b*- were prepared for calculating $\overline{\chi}_{ab}$. As mentioned above, all the *a*-*b* interactions were assumed to be equivalent to the average interaction parameter $\overline{\chi}_{ab}$ to reduce the number of interaction energy terms, and thus the model segments -*a*- and -*b*- were produced separately from the other model segments. No segments are allowed to make contact in the chain direction due to the chain connectivity. Thus, all segments were made inaccessible on both sides in the chain direction by introducing two dummy atoms at the head and tail positions of the segments, which prevented contacts with other atoms.

A particular configuration of segments i and j was produced with them touching each other, and the pairwise interaction energy for the configuration was calculated. The same procedure was repeated 100,000 times, and the Metropolis criterion was used to determine whether to accept these configurations at a certain temperature. The interaction energy $w_{kl}(T)$ was determined by averaging the energies of all accepted configurations at a given temperature. The coordination numbers z_{kl}s were obtained by calculating and averaging the possible numbers of nearest neighbors (l) in contact with the center segment (k) over 500 trials. More detailed explanations can be found elsewhere [27].

The miscibility of PS/PVME blends are investigated in the following procedure. An atactic PS chain with 20 repeating units and 50% of the meso diad fraction is produced, and then packed into a cubic simulation box under periodic boundary conditions, where the torsional angles are randomly created. The initial structure is roughly minimized and then relaxed by performing NVT-MD at 1000 K for 100,000 steps (1 step corresponds to 1 fs) to overcome the local minimum energy barrier. The structure is equilibrated by NpT-MD at 400 K for 100,000 steps and then NpT-MD at 300 K for 100,000 steps, where the coordinates are stored at every 200 steps for the confirmation of the equilibration and the analysis. The equilibration is monitored by the total energy and the volume of the simulation box, because these two quantities are most important in this study. The six equilibrated structures extracted from the MD trajectory at 300 K are energy-minimized for both the coordinates of the atoms and the cell parameters [30]. After the minimization is converged, a short time NVT-MD begins at high temperature. This minimization procedure followed by a short time NVT-MD is repeated three times. After the final minimization, the close-packed state at 0 K is obtained. It is reported that the energy-minimization results are almost independent of the initial temperature of the model polymer [26, 30]. Thus, the initial models for energy-minimization are prepared only at 300 K in this simulation. The same procedures are also applied to the PVME. In summary, the volume-temperature and energy-temperature data are calculated from the results of the NpT-MD, and the close-packed state at 0 K is directly obtained from the energy minimization. The characteristic parameters are then evaluated by a simple analysis of the simulation results.

2.3
Results and Discussion

2.3.1
Homopolymer/Copolymer Blends

Total charge values for the segments *aaa*, *aab*, and *bab* produced by the charge equilibration method are –0.009, –0.111, and –0.204, respectively. It is recalled that the total charge values of the segments *aaa*, *aab*, and *bab* correspond to those of the center monomer *a* only in each triad. The electron density in *a* increases with the substitution of *b* which donates electrons more readily than *a*. Total charge values of the segments *aba*, *abb*, and *bbb* are 0.181, 0.028, and 0.032. This means that a monomer *b* connected to the electron-withdrawing monomer *a* as its neighbor has a more positive net charge than when next to a *b*. Total charge values of the other model segments, -*a*-, -*b*-, and *c* are zero.

The interaction energy parameters were calculated by the docking method for the model segments which have proper charge distributions. Figure 1a shows that the interaction between *a* and *c* is favorable, especially when *a* is activated by *b* which donates electrons to its covalent-bonded neighbors. It is noteworthy that the interaction between the sequence *aaa* and *c* is relatively unfavorable which results in a positive $\Delta\chi_a$. This means that blockiness of *a* may lead to a negative contribution to miscibility. The experimental results showed that there exist both a specific interaction between *a* and *c*, and an intramolecular repulsion between *a* and *b* [15]. The experimental trends are seen in Fig. 1b, where $\overline{\chi}_{ab}$ is seen to be strongly positive whereas $\overline{\chi}_{bc}$ has a strongly negative value. Balazs et al. [16] assumed that $\overline{\chi}_{ac} = \chi_{aab:c} = \chi_{baa:c} = \chi_{bab:c} \neq \chi_{aaa:c}$ and that $\overline{\chi}_{bc} = \chi_{abb:c} = \chi_{bba:c} = \chi_{aba:c} \neq \chi_{bbb:c}$ to reduce the number of interaction terms. However, in this simulation, all the interaction terms such as $\chi_{aab:c}(=\chi_{baa:c})$, $\chi_{bab:c}, \chi_{aaa:c}, \chi_{abb:c}\ (=\chi_{bba:c}), \chi_{aba:c}$, and $\chi_{bbb:c}$ were calculated, and $\overline{\chi}_{ac}$ and $\overline{\chi}_{bc}$ were taken as average values, because it is unreasonable to assume $\chi_{aab:c} = \chi_{bab:c}$ and $\chi_{abb:c} = \chi_{aba:c}$ as one sees from Fig. 1a. Cantow and Schulz [31] also criticized Balazs et al.'s assumption. They asserted that the assumption cause some inconsistency. They redefined $\Delta\chi_a$ and $\Delta\chi_b$ as $\Delta\chi_a = \chi_{aaa:c} - \chi_{aab:c} = \chi_{aab:c} - \chi_{bab:c}$ and $\Delta\chi_b = \chi_{bbb:c} - \chi_{bba:c} = \chi_{bba:c} - \chi_{aba:c}$. Figure 1 shows $\Delta\chi_a \approx \overline{\chi}_{ac}$ and $\chi_{aaa:c} - \chi_{aab:c} \approx \chi_{aab:c} - \chi_{bab:c}$ at above room temperature, and $\Delta\chi_a$ in this simulation has almost the same value as the $\Delta\chi_a$ obtained using the definition of Cantow et al. In the case of $\Delta\chi_b$, the situation is not the same as $\Delta\chi_a$, but the $\Delta\chi_b$ has too small value compared with $\Delta\chi_a$ to affect χ_{dist}. Consequently, results for χ_{dist} are well consistent with those obtained from the definition of Cantow et al.

The total interaction energy parameter χ_{tot}, the composition dependent component χ_{comp}, and the sequence distribution dependent component χ_{dist} were obtained by incorporating the simulated interaction parameters into Eqs. (3)–(5). The monomer fractions (f_a and f_b) and diad fraction (f_{ab}) were determined by varying θ and δ in Eqs. (8) and (9). An increase of AA content in SAA produces more AA-EO interactions which is favorable for miscibility and thus leads to a negative χ_{comp} (Fig. 2a). In the limit that the composition of AA approaches 1.0, χ_{dist} becomes equal to $\Delta\chi_a$. In the other limit that the composition of AA approaches 0.0, χ_{dist} becomes equal to $\Delta\chi_b$. Because $\Delta\chi_a$ is larger than $\Delta\chi_b$, χ_{dist} increases with the AA content (Fig. 2b). The blockiness of the sequence represented by lower θ has a negative effect on the miscibility, which seems to arise from the strong tendency of $\chi_{aaa:c} > \chi_{aab:c} > \chi_{bab:c}$ as shown in Fig. 1a. As a result, the PEO is more miscible with the SAA copolymer having alternating sequence (Fig. 2c).

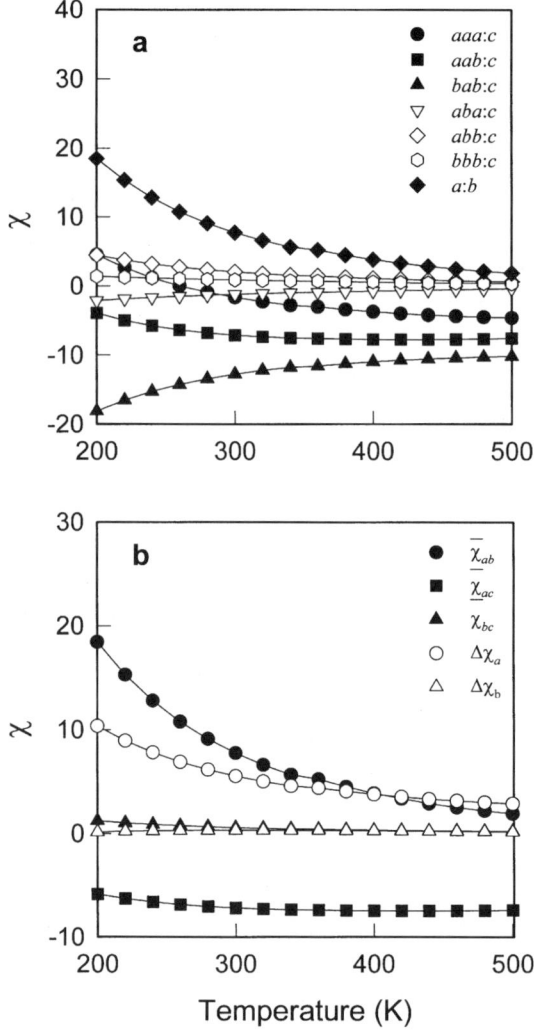

Fig. 1a,b. Interaction energy parameters as a function of temperature obtained by atomistic modeling [22]

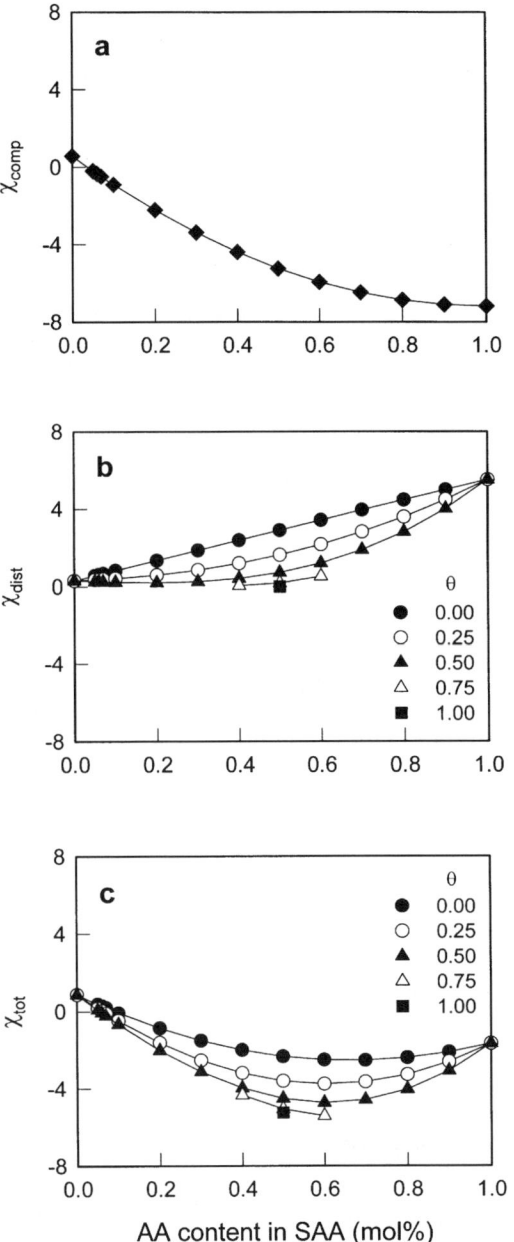

Fig. 2a–c. Interaction energy parameters at room temperature obtained by combining atomistic modeling and the theory of Balazs et al.: **a** the composition dependent interaction energy parameter χ_{comp}; **b** the sequence distribution dependent interaction energy parameter χ_{dist}; **c** the total interaction energy parameter χ_{tot} [22]

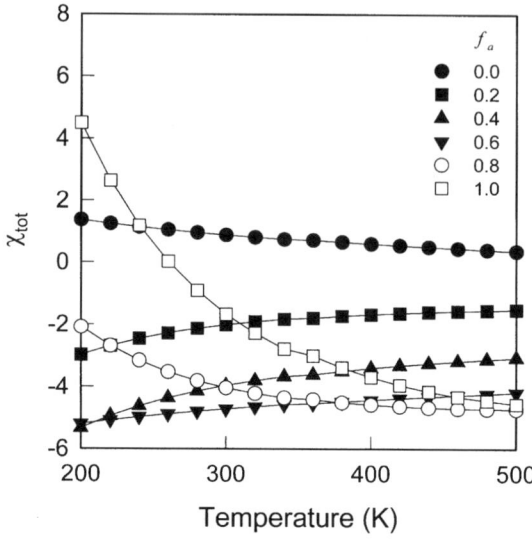

Fig. 3. Total interaction energy parameter as a function of temperature for $\theta = 0.5$ [22]

The temperature dependence of the total interaction parameter shows that there exists an optimum condition for the composition at a given temperature (Fig. 3). Binary blends of PEO/PS and PEO/PAA are immiscible and miscible, respectively, at room temperature. The shape of curves implies that the homopolymer/homopolymer blends will exhibit UCST behaviors. A drastic effect of the sequence distribution on the miscibility can be found in Fig. 4. As the AA content in SAA increases from 5 mol% (Fig. 4a) to 7 mol% (Fig. 4b) to 10 mol% (Fig. 4c), the blend becomes more miscible. The blend with random copolymers becomes miscible at a composition between 5 and 7 mol%, which agrees well with the experimental results [15]. At 7 mol%, the blend with block copolymers shows positive χ, while the blend with random copolymers has negative χ. This is very interesting because the miscibility could be controlled only by the change of copolymer sequence distributions.

2.3.2
Homopolymer/Homopolymer Blends

Since v_{sp}^* is defined as the specific volume at close-packed state and p^* is equal to ε^*/v^*, i.e., the cohesive energy density at close-packed state [17], the specific volume at 0 K corresponds to v_{sp}^*, and the cohesive energy density at 0 K to p^*. The T^* is obtained by inserting the values of p^*, v_{sp}^*, and simulated (T, v_{sp}) data at room temperature into the lattice fluid theory. The absolute values of simulated equation-of-state parameters may not be the same as the experimental ones as shown in Table 1, because the procedures obtaining the parameters are differ-

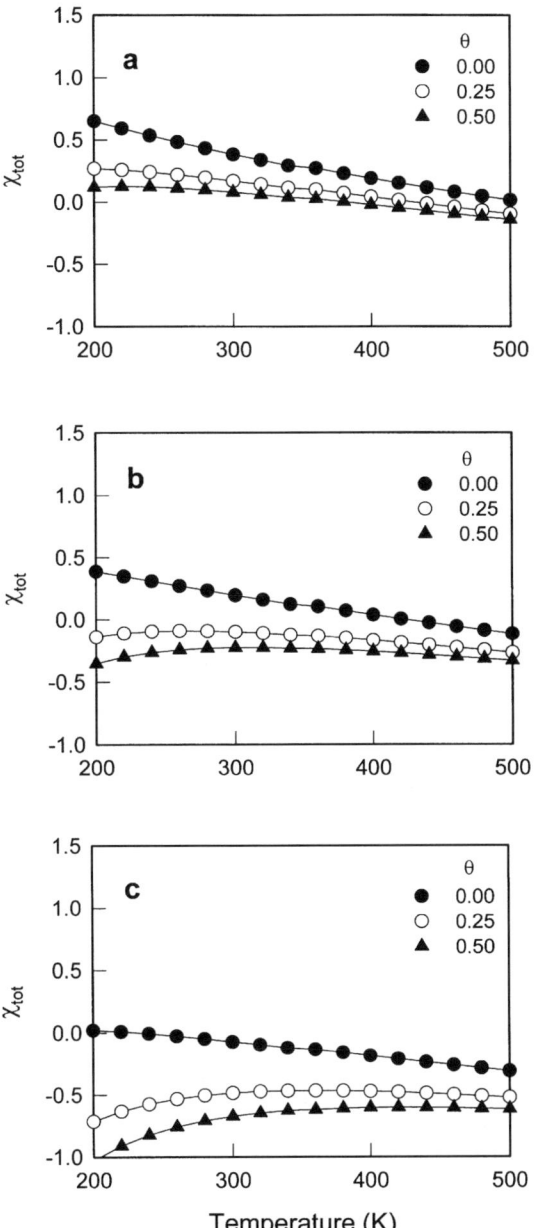

Fig. 4a–c. Total interaction energy parameter between PEO and SAA for: **a** 5 mol%; **b** 7 mol%; **c** 10 mol% of AA in SAA [22]

Table 1. Equation-of-state parameters [23]

		p^* (MPa)	v^*_{sp} (cm^3 g^{-1})	T^* (K)
PS	simul.	415 ± 6	0.927 ± 0.006	676
	exp.[a]	357	0.905	735
PVME	simul.	446 ± 18	0.930 ± 0.007	639
	exp.[b]	363	0.909	657

[a] From [32]
[b] From [33]

ent from each other. They are, however, comparable to experimentally determined values. The relative magnitude of the characteristic parameters for two polymers has been well produced in the simulation, which is very important for determining the shape of the phase diagram and the qualitative trend of surface tension.

Figure 5 shows the temperature dependence of the surface tension. The differences between calculated values and the experimental ones do not exceed ca. 1 mN m^{-1}. An adjustable parameter is not used by assuming that the $\tilde{\kappa}$ does not vary with the temperature and is fixed at 0.5, a theoretical value for both PS and PVME. This indicates that the simulated equation-of-state parameters for the component polymers are reasonable. It has been known that the LCST behaviors are originated from the specific interactions between components and/or the finite compressibility of mixture and that the phase separation is entropically

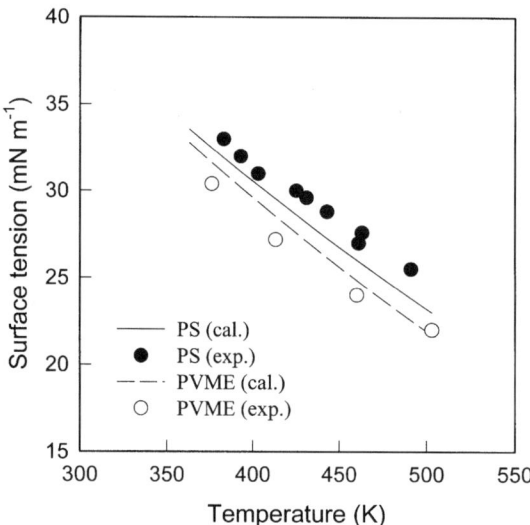

Fig. 5. Temperature dependence of surface tension of PS and PVME [23]. The *circles* represent the experimental data [34], and the *lines* are the calculated values using the simulated equation-of-state parameters. The dimensionless constant $\tilde{\kappa}$ in Eq. (13) is fixed at a theoretical value of 0.5 such that there is no adjustable parameter

driven in either case. Determination of the interaction parameter between PS and PVME is essential to predict the phase diagram using the equation-of-state theory. The interaction between PS and PVME can be represented by a dimensionless parameter ζ in Eq. (18), but this mixing rule can be applied only to the situation where the dispersion force is dominant, that is, there is no remarkable specific interaction. Although the evidence of the specific interactions between PS and PVME has been reported [32, 35, 36], it has been presumed that there is only weak specific interaction between PS and PVME, and the Flory's equation-of-state theory, in which specific interactions are not taken account, could be applied to the system [35]. In this work, the extent of the specific interaction is also assumed to be weak enough for the original lattice fluid theory to be applied. The parameter ζ might be simulated by the so-called docking method [22, 27, 28] where each ε_i (i = PS, PVME, or PS:PVME) is calculated and then one can determine the value of ζ directly from Eq. (9). However, ζ is too sensitive to extract an exact value from the method without any flexibility. Thus, the critical temperature in the phase diagram is fitted to the experimental one by varying ζ (Fig. 6). The best fit is obtained when $\zeta = 1.00066$. The shape of calculated spinodal agrees well with the experimental one, indicating that the determination of the equation-of-state parameters from molecular simulations and some assumptions adopted here are reasonable.

The theoretical treatment accounting for nonrandom mixing which may be induced by the specific interactions was first carried out by Guggenheim [38]. Sanchez and Balazs [39] have generalized the lattice fluid model by introducing the idea of specific interaction in an incompressible binary blend into the origi-

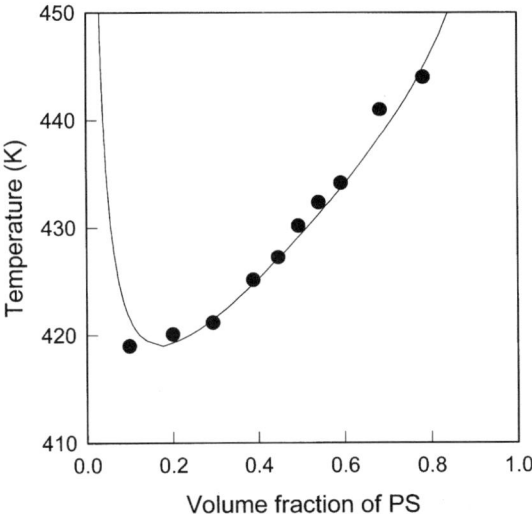

Fig. 6. Phase diagram of PS/PVME blend [23] where the line represents the calculated results using the simulated equation-of-state parameters with $\zeta = 1.00066$ and *filled circles* represent experimental spinodal data [37]

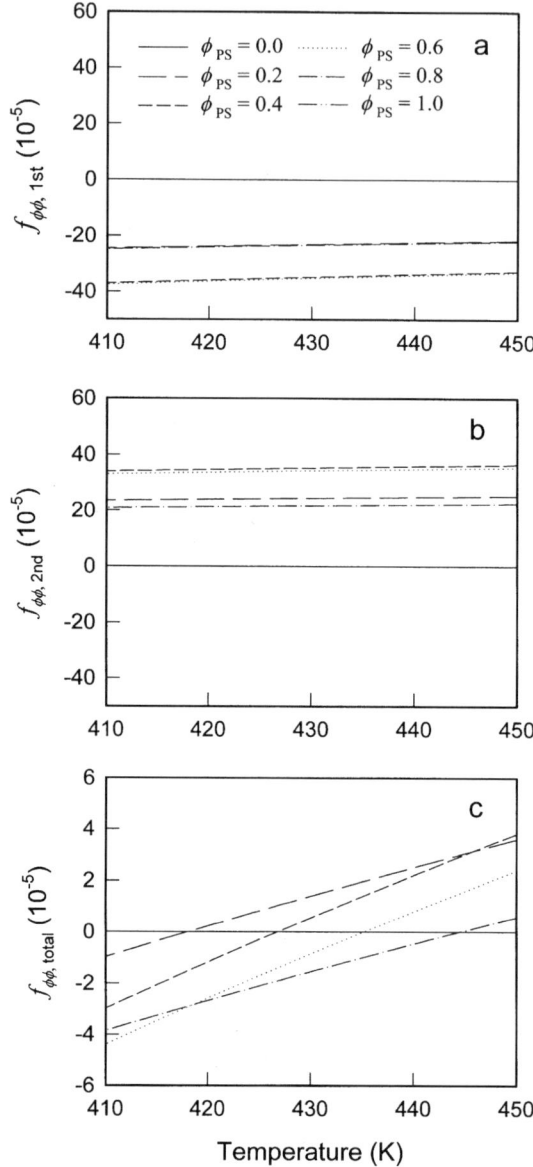

Fig. 7a–c. The second derivative of the free energy as a function of temperature [23]: **a** the first term in Eq. (16); **b** the second term; **c** the total

nal lattice fluid theory. However, the generalized lattice fluid theory has more parameters to be fitted than the original theory. It is very significant that the spinodal curve could be successfully predicted by using only one adjustable parameter when simulated values of equation-of-state parameters are used, con-

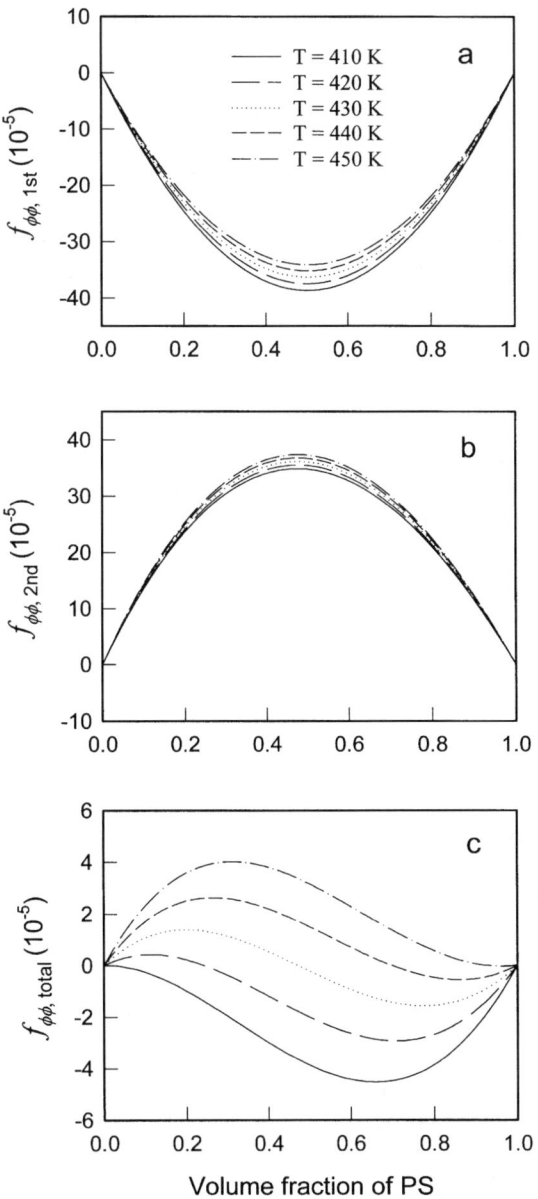

Fig. 8a–c. The second derivative of the free energy as a function of composition [23]: **a** the first term in Eq. (16); **b** the second term; **c** the total

sidering that the original lattice fluid theory predicts a very narrow spinodal curve when experimental values of equation-of-state parameters are used with one adjustable parameter [39].

When some conditions such as infinite molecular weight, near atmospheric pressure and $v^*_{PS} = v^*_{PVME}$ are assumed, the simplified equation-of-state in Eq. (16) leads to an easy interpretation on the phase diagram. Figures 7 and 8 show the components of the second derivative of the free energy per mer in Eq. (16) as a function of temperature and composition, respectively. The first term in Eq. (16), the enthalpic contribution slightly decreases in magnitude with the temperature and is symmetric with respect to the volume fraction of the PS. The second term, the compressibility contribution slightly increases with the temperature and is asymmetric with respect to the volume fraction of PS. This asymmetry leads to the asymmetric shape of the phase diagram, where the critical point for a LCST is rich in PVME with the lower T^*. The total sum results in the immiscibility at higher temperatures and PVME-rich region. In this system, both the compressibility and the specific interaction shown by the negative interaction parameter play important roles in developing the LCST behavior.

3
Block Copolymer as a Compatibilizer

The mechanism and dynamics of phase separation in polymer blends have long been the subject for many researchers in an attempt to obtain high-performance materials by controlling morphology [40–57]. Recently, phase separation behavior in polymer blends containing a small amount of block copolymer has been investigated [58–64], since a block copolymer is quite often used as a compatibilizer in many polymer blends for miscibility enhancement between immiscible polymer pairs, leading to improved mechanical properties [65–67]. Roe and coworkers [58, 59] have investigated the effects of butadiene-styrene block copolymer (PB-b-PS) on the kinetics of phase separation of off-critical mixtures of low molecular weight PB and PS mainly in the late stage of phase separation. They found that the growth rate of the average particle size of the dispersed phase is reduced upon adding a small amount of the PB-b-PS due to the reduction of the interfacial tension between the demixed phases, resulting from the accumulation of block copolymer at the interface. Hashimoto and Izumitani [60, 61] have studied the dynamics of phase separation for near-critical mixtures composed of PB and poly(styrene-ran-butadiene) (SBR) to which a small amount of SBR-b-PB is added. They observed that block copolymers added to immiscible blends slow down the phase-separation process, and that the higher the concentration of block copolymers in the blend, the slower the dynamics. Laradji et al. [62] performed a computer simulation for the dynamics of the phase-separation process of binary mixtures containing surfactants, based on a time-dependent Ginzburg-Landau model, where the surfactant molecule has an amphiphilic nature, i.e., a surfactant molecule in an A/B binary mixture consists of an A-philic part and a B-philic part. Kawakatsu et al. [63] have studied the effects of added surfactant on phase separation dynamics using the hybrid model. In both cases, it was found that the coarsening rate of domains is considerably decreased in the late stage due to the reduced interfacial tension of the surfactant-absorbed interface. On the other hand, Lin et al. [64] found that, based on the value of the Onsager coefficient during the early stage and the time taken for the sample to enter the late stage of spinodal decomposition (SD), the addition of block copolymer leads to an acceleration of SD in the system of polymethylbutylene (PM)/polyethylbutylene (PE)/PM-b-PE, although the addition of block copolymer results in a finer structure over the entire time range. In this case, the block copolymer was not located preferentially at the interface, but uniformly distributed in the sample, unlike what is usually observed in mixtures of two homopolymers and a block copolymer. In this section, the effects of interaction energy, chain length, and composition of C-b-D block copolymers on the phase separation dynamics and mechanism of A/B immiscible homopolymer blend are systematically analyzed using Monte Carlo simulation [68–71].

3.1
Simulation Methods

The simulations are performed on a simple cubic lattice of 50×50×50 sites with periodic boundary conditions. An A/B binary blend system is first simulated and then A/B/C-b-D ternary blend systems containing various types of block copolymers, where all polymer chains are represented as self-avoiding and mutually avoiding near-neighbor walks on the lattice are simulated. The composition of homopolymers A and B is fixed at 75/25 v/v in all cases, and the homopolymer chains have the same chain length, i.e., $N_A = N_B = 10$. The phase separation of homopolymer chains A and B is induced by introducing a positive pairwise interaction energy $\varepsilon_{AB}(= \varepsilon'_{AB}/k_BT$, where k_B is the Boltzmann constant and T is temperature). In all cases, as ε_{AB} has a constant positive value (+0.5), homopolymers A and B are phase-separated. The blocks of added copolymer have chemically different structures from the homopolymers, and thus there are a total of six different interaction energy parameters including ε_{AB}. All interaction energies used in the blend systems and the nomenclature of block copolymer added in this work are summarized in Table 2. The total chain length N_{block}, the composition $f (= N_C/N_{block}$, where N_C is the length of block C), and the interaction parameters ε of block copolymers are systematically varied in order to investigate their effects on the phase separation process of immiscible polymer blends. In all cases, the amount of added block copolymers is fixed at 5.7% of the total sites occupied by all chain segments. The total volume fraction of polymers is set at 0.61.

The bond fluctuation model [72] is used to simulate the motion of the polymer chains on the lattice. In this model, each segment occupies eight lattice sites of a unit cell, and each site can be a part of only one segment (self-avoiding walk condition). This condition is necessary to account for the excluded volume of the polymer chains. For a given chain, the bond length between two successive seg-

Table 2. Classification of block copolymers according to their interaction type and composition

Block copolymer	ε_{AB}	ε_{CA}	ε_{CB}	ε_{DA}	ε_{DB}	ε_{CD}	Composition f
Repulsive block copolymer							
R11f5	0.5	0.1	1.0	1.0	0.1	1.0	0.5
R21f5	0.5	0.1	2.0	2.0	0.1	1.0	0.5
R22f5	0.5	0.1	2.0	2.0	0.1	2.0	0.5
R22f7	0.5	0.1	2.0	2.0	0.1	2.0	0.7
Attractive block copolymer							
A01f5	0.5	−0.1	0.5	0.5	−0.1	0.5	0.5
A02f5	0.5	−0.2	0.5	0.5	−0.2	0.5	0.5
A05f5	0.5	−0.5	0.5	0.5	−0.5	0.5	0.5
A05f7	0.5	−0.5	0.5	0.5	−0.5	0.5	0.7

ments can have all possible lengths from 2 to $10^{1/2}$, restricted to a particular set of vectors [73]. For a cubic lattice, the possible vectors can be derived from the basic set {(2,0,0), (2,1,0), (2,1,1), (2,2,1), (3,0,0), and (3,1,0)}. This set of bonds guarantees the self-avoiding condition and recognizes the non-intersection rule in the course of their motion. The polymer chains move by random diffusion jumps of their segments to nearest-neighbor lattice sites simultaneously obeying the excluded volume and bond-length constraints.

To monitor the time evolution of the long-range ordering, the collective structure factor at the constant time interval of phase separation is calculated by Eq. (20):

$$S(\mathbf{q},t) = \langle \sum \exp(i\mathbf{q} \cdot \mathbf{r}) \phi(\mathbf{r}+\mathbf{r}',t) \phi(\mathbf{r}',t) - \langle \phi \rangle^2 \rangle / L^3 \qquad (20)$$

where $\mathbf{q} = (2/L)(x\mathbf{i} + y\mathbf{j} + z\mathbf{k})$ with $x, y, z = 1,2,3,\ldots,L$, where L is the dimension of the system, and the angular brackets denote an average over the number of runs. This quantity represents a Fourier transform of pair-correlation functions and is spherically averaged to smooth the results as follows:

$$S(q,t) = \sum_{q}{}' S(q,t) / \sum_{q}{}' 1 \qquad (21)$$

where $q = (2\pi/L)n$ with $n = 1,2,\ldots,$ and each sum $\sum'$ for a given value of n is over a spherical shell defined by

$$n - \frac{1}{2} \leq \frac{qL}{2\pi} \leq n + \frac{1}{2} \qquad (22)$$

The first moment $q_1(t)$ of the structure factor is also computed in order to observe the coarsening processes in the later stage of phase separation more clearly. This quantity $q_1(t)$, the inverse of which is a measure of the average domain size, is defines as

$$q_1(t) = \frac{\sum_q q S(q,t)}{\sum_q S(q,t)} \tag{23}$$

For typical measure of domain size, one usually considers either the location $q_{max}(t)$ of the peak of the spherically averaged structure factor or some moment of $S(q,t)$. Here only the first moment q_1 will be considered as a measure of average domain size, since q_{max} cannot be precisely determined due to the discretization of the scattering vector **q** in a finite lattice. Five independent runs are performed for each case, and all the results are reported by averaging the data from five independent runs.

3.2
Results and Discussion

3.2.1
A/B Binary Blends

The simulation result for the time evolution of structure factors as a function of the scattering vector q for an A/B 75/25 (v/v) binary blend is shown in Fig. 9 where time elapses in order of Fig. 9c to 9a. The structure factor $S(q,t)$ develops a peak shortly after the onset of phase separation, and thereafter the intensity of the peak S_{max} increases with time while the peak position q_{max} shifts toward smaller values with the phase-separation time. This behavior suggests that the phase separation proceeds with evolution of periodic concentration fluctuation due to the spinodal decomposition and its coarsening processes occurring in the later stage of phase separation. These results, consistent with those observed in real polymer mixtures, indicate that the simulation model can reasonably describe the phase separation process of real systems.

The process of phase separation is also accompanied with a change in the number of nearest neighbor contacts and the linear dimensions of the chains, as shown in Fig. 10. Figure 10a shows the variation of the number of contacts per chain between the same kind of segments (A-A) on different chains and between the different kind of segments (A-B) with the phase separation time. The number of contacts between the same kind of segments, n_{AA}, increases fast and then levels off, while the number of contacts between the different kind of segments, n_{AB}, decreases quickly in the early stage of phase separation and then levels off as the phase separation further proceeds. The variation of the radius of gyration of chains with time is shown in Fig. 10b. In both cases of chains A and B, the contraction of coils occurs at the early stage of phase separation, the extent of which is larger for chain B corresponding to the minor phase than for chain A corresponding to the major phase.

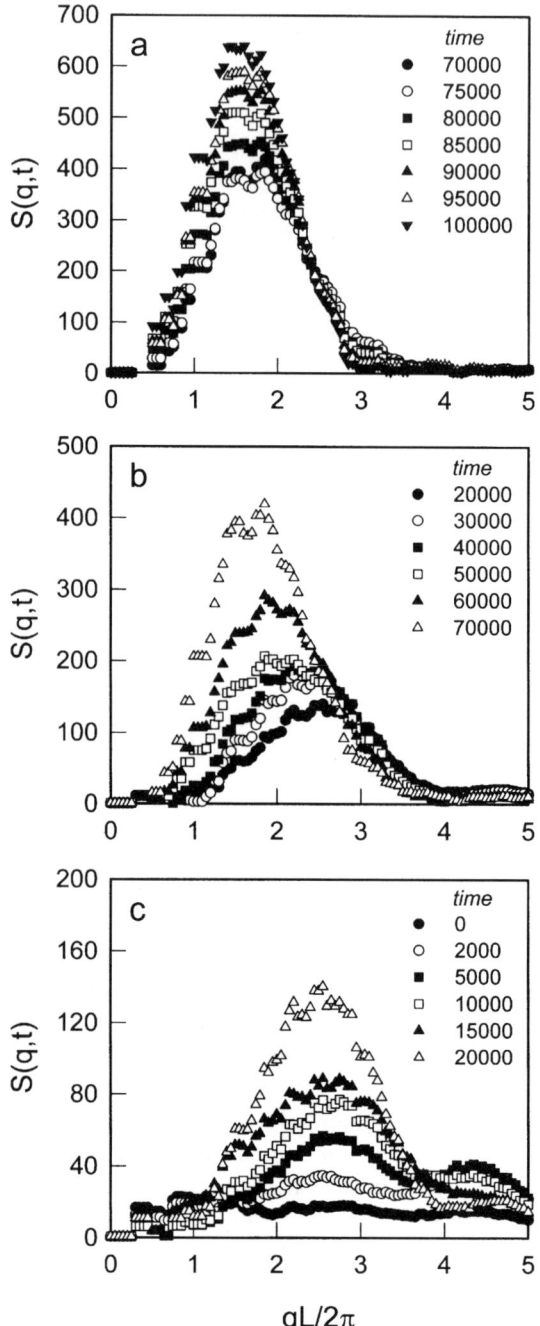

Fig. 9a-c. Time evolution of structure factor during phase separation for an A/B binary blend [71]. Time elapses in order of **c** to **a**

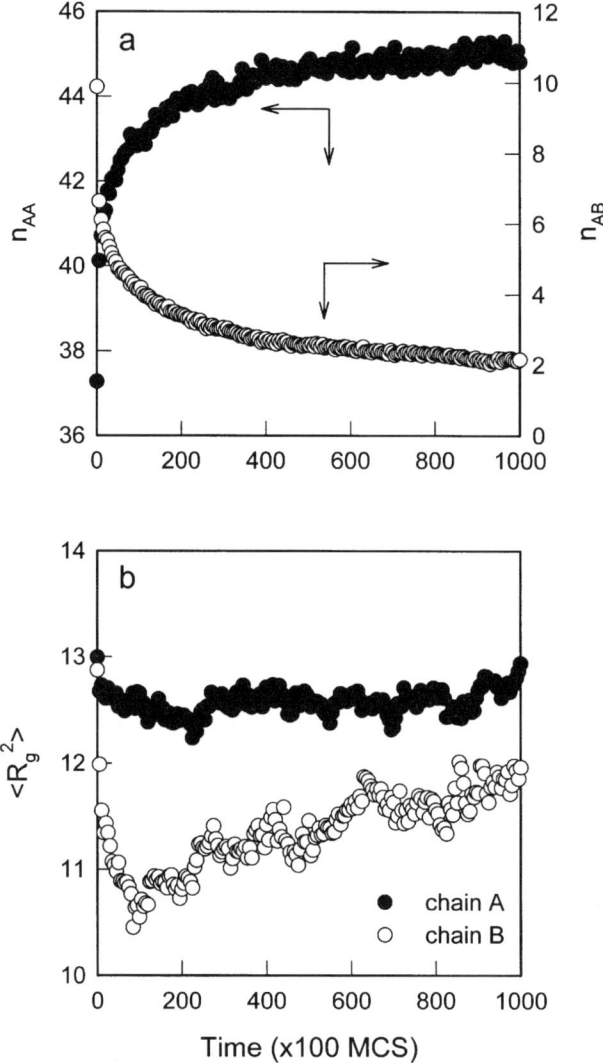

Fig. 10a,b. Change of: **a** the number of nearest neighbor contacts; **b** the linear dimension of chain A and B with time for A/B binary blend [68]

Generally, the spinodal decomposition of polymer mixtures is classified into three stages, each of which is called early, intermediate, and late stage, respectively [50]. In the early stage of spinodal decomposition, whose dynamics can be well described by the linearized theory [74], the amplitude of the fluctuations exponentially increases with time without any variation in the wavelength of the fluctuations. The phase separation up to 5000 Monte Carlo (MC) steps in Fig. 9c corresponds to this stage. With increasing amplitude the linear approximation in the linearized theory fails, indicating the onset of the intermediate stage. In this stage, both the wavelength and the amplitude of the concentration fluctuation grow with time. In this simulation, the profiles up to 40,000 MC steps in Fig. 9b correspond to the intermediate stage. In the late stage of spinodal decomposition, the amplitude of fluctuations reaches an equilibrium but the wavelength still increases with time.

The application of the Cahn's linearized theory to polymer mixtures was first developed by de Gennes [40] and Pincus [41], based on the Flory-Huggins free energy expression. Subsequently, Binder [42] incorporated the concept of thermal noise, introduced earlier by Cook [75], into the polymer-based derivation. This theory implies an exponential growth of structure with time:

$$S(q,t) = S_x(q) + [S(q,0) - S_x(q)]\exp(2R(q)t) \qquad (24)$$

where $S(q,t) \sim I(q,t)$ is the collective structure factor, $S_x(q)$ represents a virtual structure factor, and $R(q)$ is the growth rate for the q-Fourier mode of the fluctuation determined by the collective diffusivity D_{app}, as given by

$$R(q) = D_{app}q^2\left[1 - \frac{1}{2}\left(\frac{q}{q_m(0)}\right)^2\right] \qquad (25)$$

However, it was not possible to find a time region in which $S(q,t)$ increases exponentially with time in the plots of $S(q,t)$ vs time t in these blend models. The structure factor increased linearly with time and then bent down, reflecting the onset of strong nonlinear effects, as in the cases observed in the simulations of spinodal decomposition in metallic alloys [76–78]. An initial stage of spinodal decomposition can be identified in Fig. 11, which shows the variation of the first

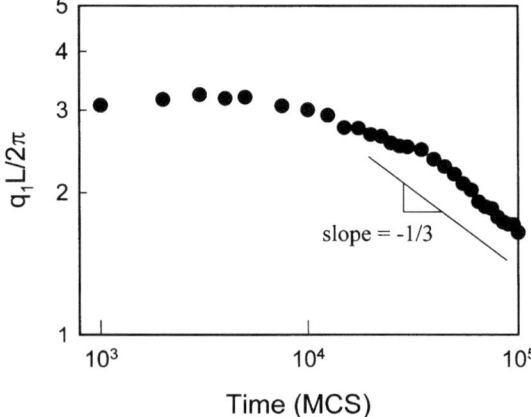

Fig. 11. Log-log plot of the first moment $q_1(t)$ of structure factor as a function of time for an A/B binary blend [71]

moment $q_1(t)$ of structure factors with time in double-logarithmic plot. In this figure, two different time regimes are clearly observed. The first regime, where q_1 is almost invariant with time, as predicted by the linearized theory, corresponds to the early stage of spinodal decomposition. After this stage, $q_1(t)$ decreases with time due to the increase in the amplitude of concentration fluctuation, indicating that the phase-separation process moves to the intermediate stage. It is also observed in Fig. 11 that $q_1(t)$ is proportional to $t^{-1/3}$ in the late stage. The value of exponent, -1/3, is generally accepted for the spinodal decomposition in binary mixtures without the hydrodynamic interaction [79]. It should be noted that in a real polymer blend, the coarsening process proceeds with $q_1(t) \sim t^{-1}$ due to the hydrodynamic interactions [80].

In the late stage of phase separation in binary polymer mixtures, all length scales, except the microscopic ones, have the same time dependence [55, 56]. This leads to the following scaling property of structure factor:

$$S(q,t) = Mq_1(t)^{-d}F(x) \tag{26}$$

where d is the spatial dimension, $x = q/q_1(t)$, M an arbitrary normalization constant, and F the scaling function. The q_1 is used as a single time-dependent length parameter. In order to test if the scaling relation holds in this simulation, the scaled structure factor is calculated by Eq. (27):

$$F(x) = (L/\pi)q_1^3 S(q,t) / \sum_q q^2 S(q,t) \tag{27}$$

When the scaled structure factor $F(x)$ is plotted against the reduced scattering vector x for an A/B binary blend in Fig. 12, $F(x)$ becomes essentially universal

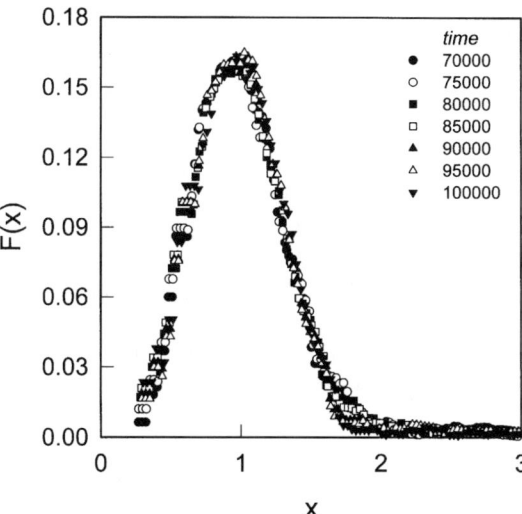

Fig. 12. Plot of the scaled structure factor $F(x)$ in the late stage of spinodal decomposition for an A/B binary blend as a function of the reduced scattering vector x [71]

with time and falls closely to a single master curve. This behavior indicates that there exists self-similarity between the phase-separated structures developed at different times in the late stage of phase separation.

3.2.2
A/B/C-b-D Ternary Blends

In general, a block copolymer added to immiscible polymer blend significantly suppresses the growth rate of phase-separated domains due to the reduction of interfacial tension resulting from a preferential localization of block copolymer at the interface. However, the retardation effect by the block copolymer is found to be dependent upon the structure of the block copolymer added, such as the interaction energy, the chain length, and the composition of block copolymer.

3.2.2.1
Effects of the Interaction Energy of Block Copolymers

Figure 13 shows the time evolution of structure factors during phase separation for an A/B binary blend and two A/B/C-b-D ternary blends containing a small amount of block copolymers that have the different types of interaction energies between the blocks and the homopolymers. It can be seen that the rate of phase separation is faster for the binary blend than for two ternary blends containing a block copolymer. This observation clearly indicates that the added block copolymer plays the role of compatibilizer. Another interesting feature from Fig. 13 is that even block copolymers having the repulsive interactions with homopolymers can suppress the growth rate of phase-separated domains, although the suppression is not so effective as the case where the block copolymer having the attractive interactions is added. This indicates that the block copolymer having repulsive interactions with the homopolymers can also act as a compatibilizer as long as block C is less compatible with homopolymer B than with homopolymer A, i.e., $\varepsilon_{CB} > \varepsilon_{CA}$ and block D is less compatible with homopolymer A than with homopolymer B, i.e., $\varepsilon_{DA} > \varepsilon_{DB}$. In this case, as reported by Vilgis and Noolandi [81], the competitive interactions of the blocks with the homopolymers force the block copolymers to preferentially localize at the interface between two phases and consequently result in the reduction in the interfacial tension. This is responsible for the retardation effect observed in Fig. 5c where the block copolymers have the positive interactions with homopolymers.

Figure 14 shows the time evolution of the structure factor during phase separation for the ternary blend systems containing attractive block copolymers with $\varepsilon_{AC}(=\varepsilon_{BD})$ being equal to –0.1, –0.2, and –0.5, respectively, where $N_{block}=12$ for all blend systems. When one compares Fig. 14a, b, and c with each other, one may realize that the retardation of phase separation by the addition of block copolymer becomes more pronounced as the interaction energies (ε_{AC} and ε_{BD}) become more negative. This phenomenon seems to be closely related to the variation of the radius of gyration of block copolymers with the magnitude of inter-

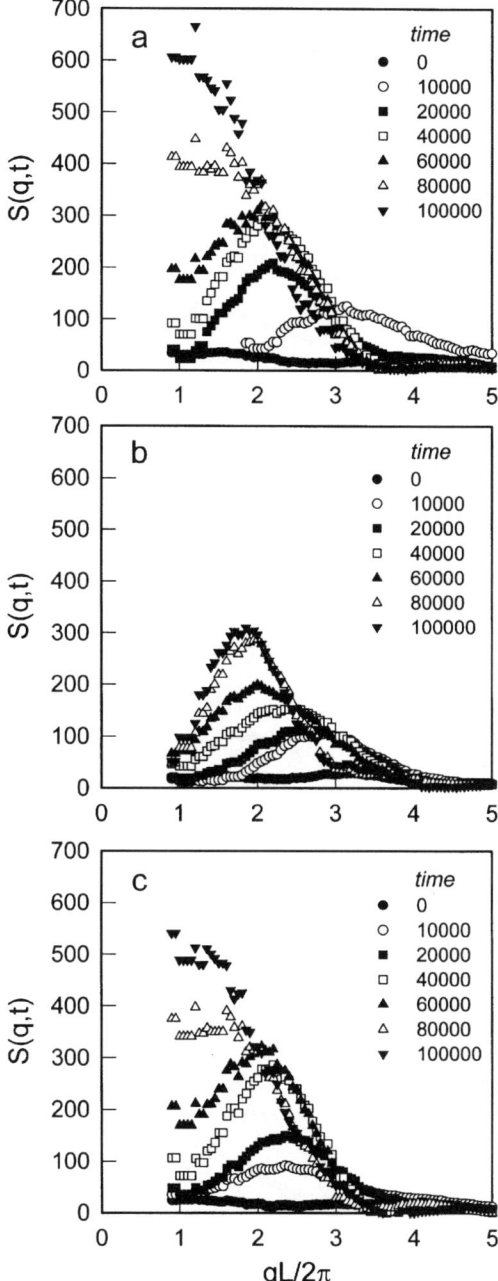

Fig. 13a–c. Time evolution of structure factor during the phase separation for: **a** a blend without block copolymer; **b** blend with block copolymers of attractive interaction (A05f5); **c** blend with block copolymers of repulsive interaction (R11f5) between the blocks and the homopolymers [70]. The total chain length of block copolymers is $N_{block} = 12$

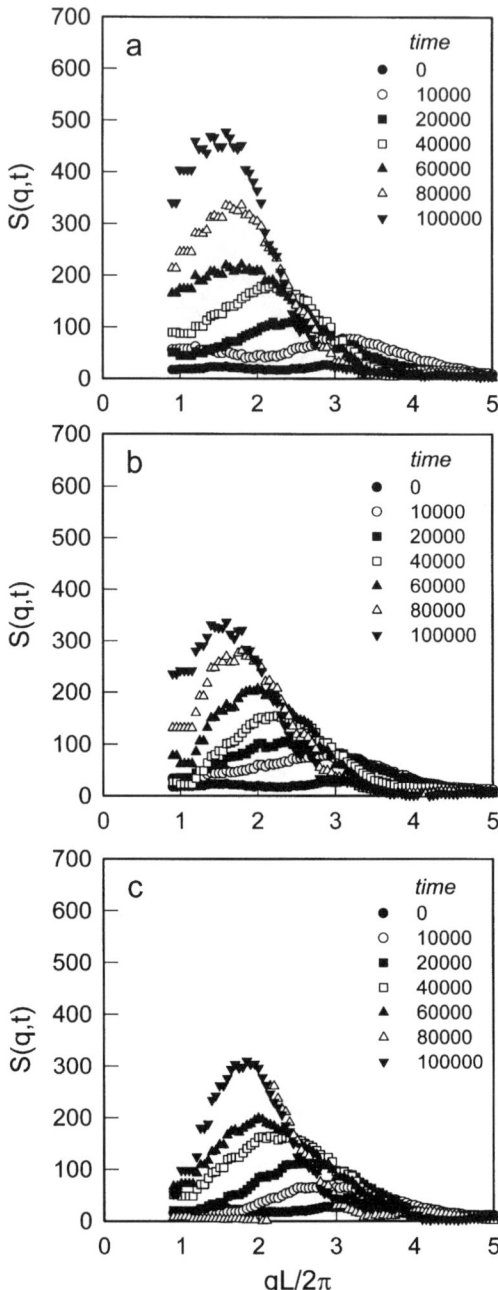

Fig. 14a–c. Time evolution of structure factor during the phase separation for ternary blends with attractive block copolymers where $N_{block} = 12$ for all blends [68]: **a** A01f5; **b** A02f5; **c** A05f5

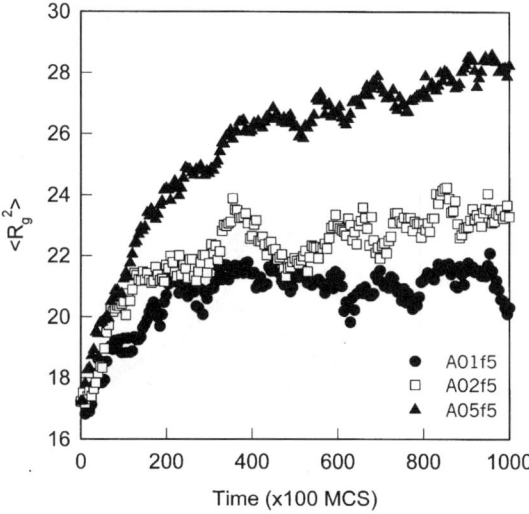

Fig. 15. Change of mean-squared radii of gyration of attractive block copolymers (N_{block} = 12) with different interaction during the phase separation [68]

action energy, as shown in Fig. 15. The block copolymers migrate to the interface between the two phases, where the blocks penetrate into their respective homopolymer phases, leading to an increase in the linear dimension of block copolymer chains. The larger the interaction between the block and corresponding homopolymer, the larger the extension of block copolymer chains becomes. Consequently, the block copolymers with more negative interactions can accumulate more at the interface, causing a larger reduction in the interfacial tension.

Figure 16 shows that the best retardation effect is obtained with R22f5 among repulsive block copolymers investigated in this study. Thus, the competitive interactions of the blocks with different homopolymers are shown to promote the retardation. In other words, as the relative repulsion between block C and homopolymer B and between block D and homopolymer A increases, i.e., if the interaction of block C (D) with homopolymer B (A) is more repulsive than that of block C (D) with homopolymer A (B), the rate of phase separation is retarded more effectively. Vilgis and Noolandi [81] also predicted strong interfacial activity of block copolymers with such interactions. Therefore, it can be concluded that the interaction energy between blocks also has an important effect on the phase separation of immiscible blends.

3.2.2.2
Effects of the Chain Length of Block Copolymers

The effects of chain length of added (attractive or repulsive) block copolymer on the phase separation dynamics of immiscible polymer blends are summarized

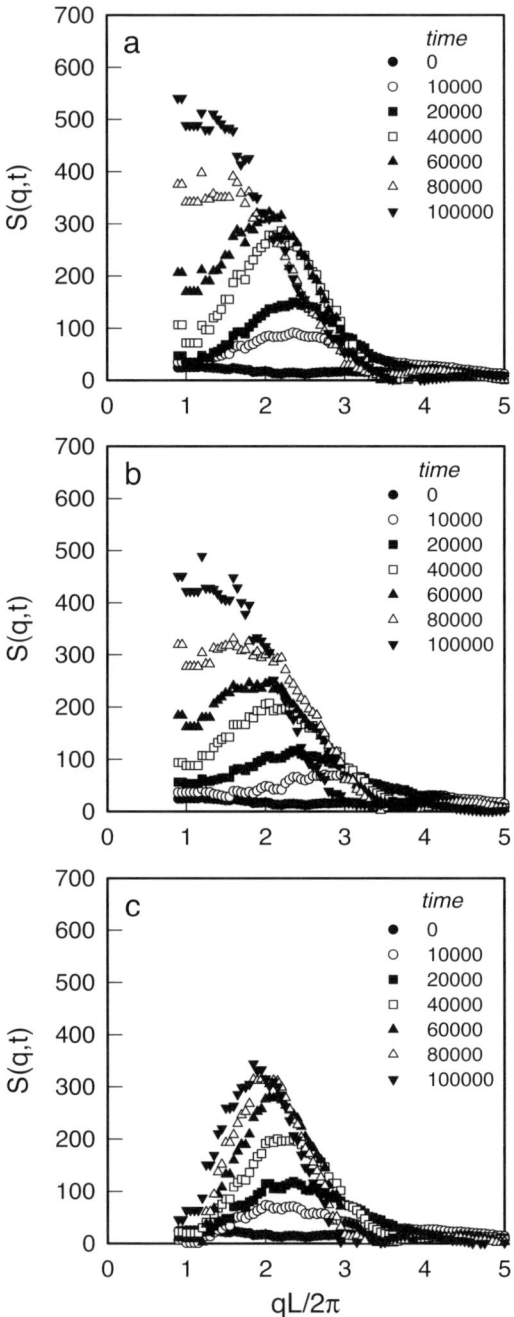

Fig. 16a–c. Time evolution of structure factor during the phase separation for ternary blends with repulsive block copolymers where $N_{block} = 12$ for all blends [70]: **a** R11f5; **b** R21f5; **c** R22f5

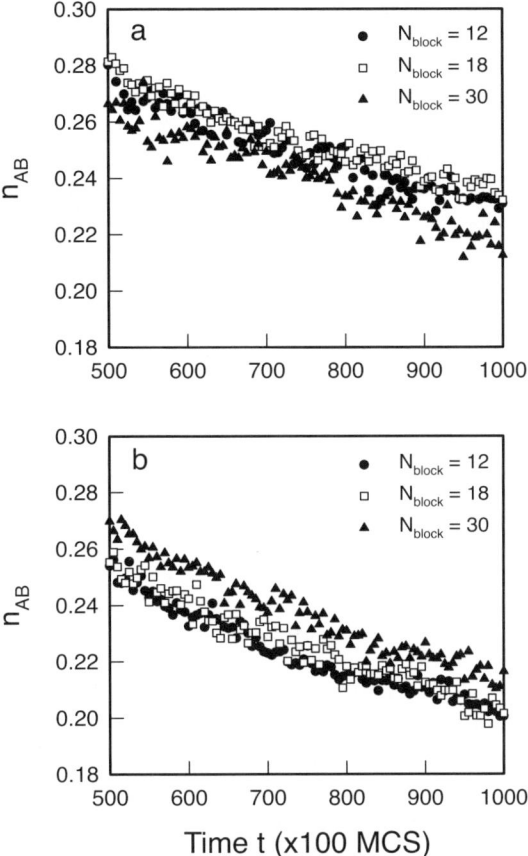

Fig. 17a,b. Change of normalized contact number n_{AB} between segments of homopolymer A and B in the ternary blends containing: **a** attractive (A05f5); **b** repulsive block copolymers (R11f5) of different chain lengths with time [71]

in Fig. 17. The contact number n_{AB} between A and B segments is a useful measure of the degree of phase separation, i.e., smaller n_{AB} means higher degree of phase separation. When one compares n_{AB} values in Fig. 17, it is found that the degree of phase separation depends on the interaction type (attractive or repulsive) and the chain length of added block copolymers.

Noolandi and coworkers [82–84] investigated the equilibrium interfacial properties for the multi-component mixtures containing a block copolymer by the numerical calculations based on the self-consistent mean field theory. For both A/B/A-b-B and A/B/C-b-D systems, they found that the interfacial tension decreases and the concentration of block copolymer at the interface increases with increasing the chain length of block copolymer. Israels et al. [85] examined the interfacial behavior of symmetric A-b-B diblock copolymers in a blend of immiscible homopolymers A and B by a similar method. They also found that the interfacial tension can be significantly reduced when the blocks in the copolymer are longer than the corresponding homopolymer. Such behavior is also observed in these simulation results. As shown in Fig. 17, as the chain length of block copolymer increases from $N_{block} = 12$ to $N_{block} = 18$, n_{AB} becomes larger over the entire time examined, indicating that the retardation effect by block copolymers is enhanced. However, when the copolymer chain length increases further to $N_{block} = 30$, a distinctly different behavior is observed in the retardation effect between the attractive and repulsive block copolymers. The retardation effect of repulsive block copolymer (Fig. 17b) becomes better when the block copolymer chain length increases further to $N_{block} = 30$, as predicted by the theoretical calculation mentioned above, whereas the retardation effect of attractive block copolymer with $N_{block} = 30$ (Fig. 17a) becomes rather worse than the cases of $N_{block} = 12$ and $N_{block} = 18$. It is known that the emulsifying effect of block copolymer is sometimes limited by the copolymer micelle formation in homopolymer phases, because this micelle makes it difficult for the system to reach an equilibrium by slowing down the migration of block copolymer to the interface. As a result, a decrease in the retardation effect of attractive block copolymer with $N_{block} = 30$ may be explained by the micelle formation of the block copolymer in homopolymer phases, although it is not possible to find a direct evidence of micelle formation in this simulation. However, the reason why the retardation of the block copolymer with $N_{block} = 30$ becomes worse could be explained by comparing the radius of gyration of block copolymers with different chain lengths. In general, as the phase separation proceeds, the block copolymer chains are extended, since they migrate to the interface during the phase separation and each block chain penetrates into its respective homopolymer phase. When Fig. 17 is compared with Fig. 18, it is seen that the degree of chain extension of block copolymer is consistent with the degree of phase separation for both types of block copolymers. The attractive block copolymer with $N_{block} = 30$ is least extended, whereas the repulsive block copolymer with $N_{block} = 30$ is most extended.

In Fig. 18, some different behavior between the attractive and repulsive block copolymer is observed. First, attractive block copolymers are more extended than repulsive ones, indicating that they basically have a better retardation effect than repulsive ones. This is a result of the fact that the blocks in repulsive block copolymers are contracted due to the repulsive interaction between blocks and homopolymers even after block copolymers migrate to the interface, although the overall chain dimension increases with time during phase separation. Another difference between the attractive and repulsive block copolymers is observed in the early stage of phase separation, where the repulsive block copolymer chains are slightly contracted. In this stage, the interface is not fully developed and most of block copolymers remain in the bulk phase. The blocks in copolymer have repulsive interactions with all homopolymers and thereby the block copolymer chains are contracted. However, as the phase separation proceeds further, the interfaces are formed and at the same time the block copolymers migrate to the interface, resulting in chain extension.

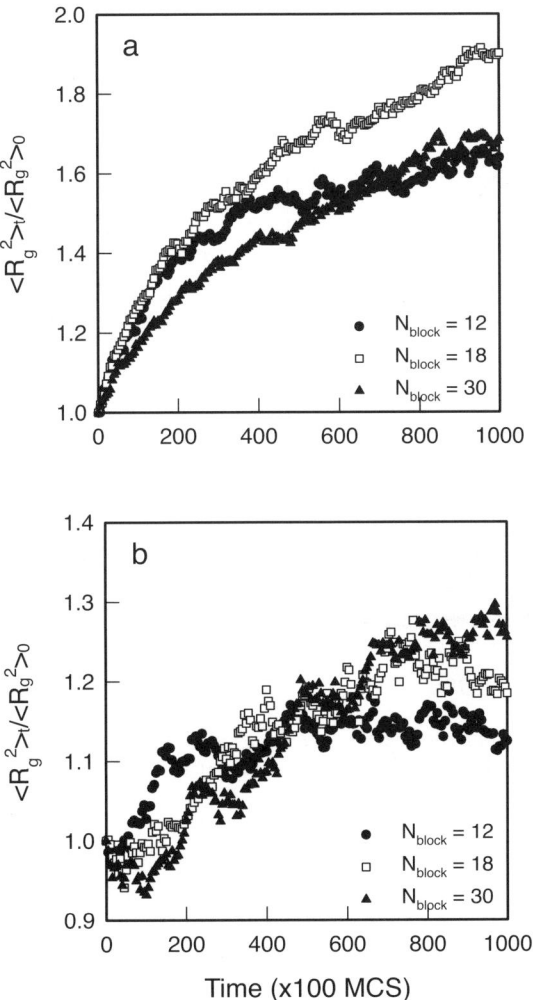

Fig. 18a,b. Change of mean squared radii-of-gyration of: **a** attractive (A05f5); **b** repulsive block copolymers (R11f5) of different chain lengths with time [71]

3.2.2.3
Effects of the Composition of Block Copolymers

For attractive block copolymers with chain length shorter than the homopolymers, the copolymer with a symmetric composition ($f = 0.5$) is more effective than the asymmetric ones in retarding the phase separation (Fig. 19a). However, for the block copolymers with chain length longer than the homopolymers, a copolymer with an asymmetric composition shows better retardation effect (Fig. 19b). The composition of a block copolymer exhibiting the best effect is de-

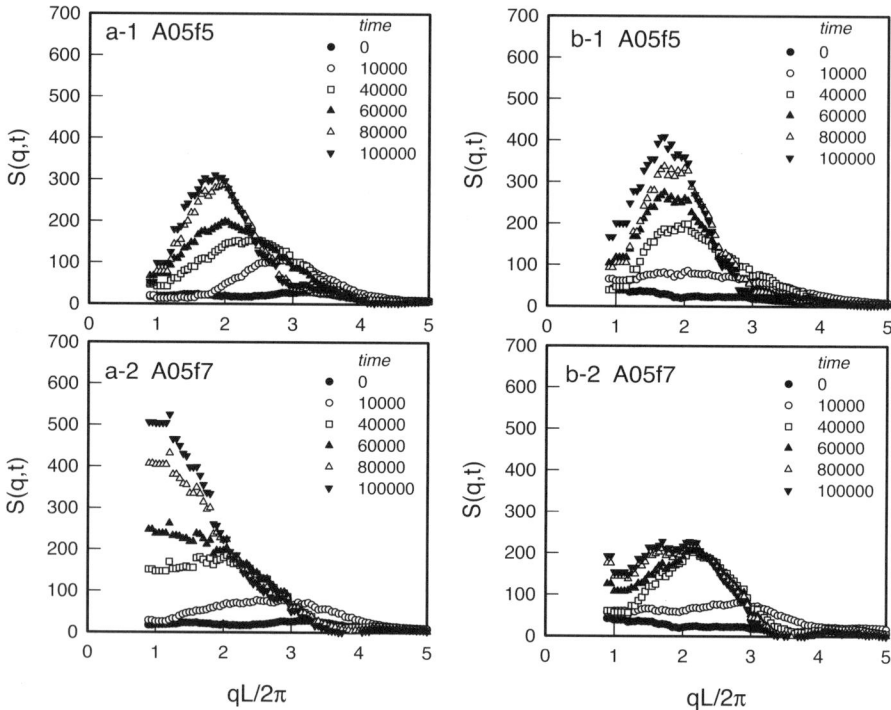

Fig. 19a,b. Time evolution of structure factor during phase separation for A/B/C-b-D ternary blends containing attractive block copolymer with **a** $N_{block} = 12$ and **b** $N_{block} = 30$ [69]

termined by the composition equal to the homopolymer blend composition because of the curvature properties of the interface, i.e., the elasticity and spontaneous radius of curvature of the interface. These results can be roughly expected by the theoretical calculation on the basis of the random phase approximation (RPA) [86] for A/B/A-b-B blend systems. Tanaka and Hashimoto [87] calculated the stability limits of the phase transitions of binary and ternary mixtures of A-b-B diblock copolymer with corresponding homopolymers A and/or B by using RPA. According to their calculation scheme, the maximum compatibilizing effect can be observed when the block copolymer having the same composition as the homopolymer blend composition is mixed in the ternary mixture. Figure 20 shows variations of the radius of gyration of the attractive block copolymers in the ternary blends during the phase separation. The extension level of block copolymer chains with two different compositions shown in Fig. 20 exactly matches the degree of the phase separation observed in Fig. 19.

However, a repulsive block copolymer exhibits different behavior in the composition effect on the phase separation of polymer blend. As observed in Fig. 21, unlike the case of attractive block copolymer, the retardation effect by the asymmetric block copolymer with $f = 0.7$ becomes rather worse than that by the sym-

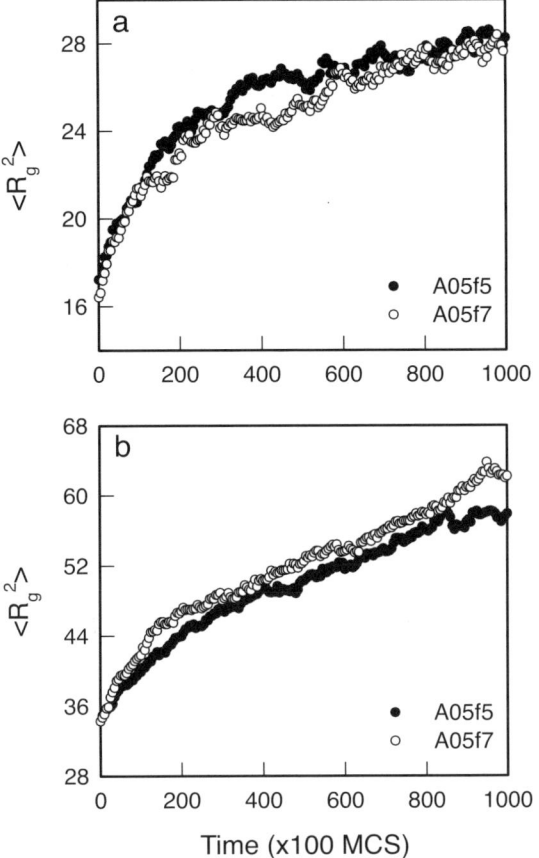

Fig. 20a,b. Change of mean-squared radii of gyration of block copolymer of chain length of: a $N=12$; b $N=30$ in A/B/C-b-D blends with different compositions [69]

metric block copolymer when the chain length of block copolymer was $N_{block} = 30$. When $N_{block} = 12$, an asymmetric one shows almost the same retardation effect as the symmetric one. The effect of the composition of repulsive block copolymer can be analyzed by considering the change of the contact number n_{AB} between A and B segments, as shown in Fig. 22. As mentioned earlier, this contact number is a measure of the degree of the phase separation. When $N_{block} = 30$ (Fig. 22b), the asymmetric block copolymer has a smaller n_{AB} value than the symmetric one over the entire time examined, indicating that its retardation effect is worse. When $N_{block} = 12$ (Fig. 22a), the asymmetric block copolymer has the similar value of n_{AB} to the symmetric one. These results are exactly consistent with ones observed in the time evolution of structure factors in Fig. 21.

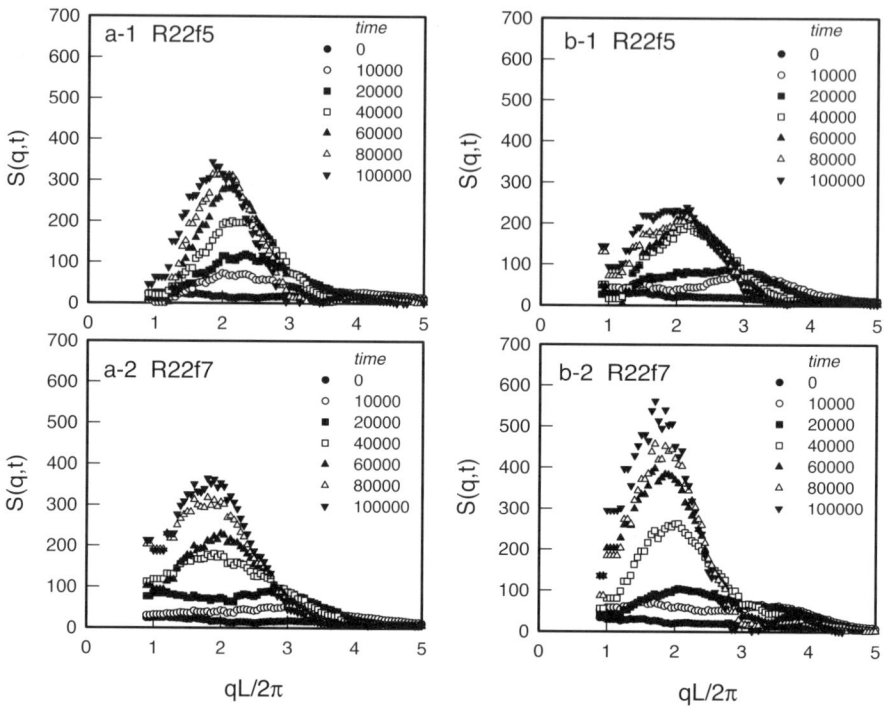

Fig. 21a,b. Time evolution of structure factor during phase separation for A/B/C-b-D ternary blends containing repulsive block copolymer with: **a** $N_{block} = 12$; **b** $N_{block} = 30$ [71]

3.2.2.4
Scaling

It has been found that the phase separation of A/B binary blends obeys the dynamic scaling law such as Eq. (27) [68]. As seen in Fig. 23, it is clear that such a scaling relation holds for ternary blend systems containing attractive block copolymers, irrespective of the chain length and interaction energy of attractive block copolymers added. The single time-dependent length parameter used for the scaling analysis was the first moment $q_1(t)$ of structure factors, the inverse of which represents the average size of the phase-separated domains. However, $F(x)$ is not universal at $x > 2$ corresponding to the scale of local domain such as the interface, implying that the interfacial structure cannot be scaled with the length parameter $q_1(t)$ characterizing the global structure. Hashimoto and coworkers [55, 56] divided the late stage of spinodal decomposition into two different stages, late stage I and late stage II. They defined the late stage I as the time interval during which the scaled structure factor $F(x)$ for high x remains unscalable with $q_1(t)$ only. Thus, in this stage, the local structures at different times do not have self-similarity, implying that local structure such as the interfacial structure would obey different scaling laws. The late stage II was defined as the

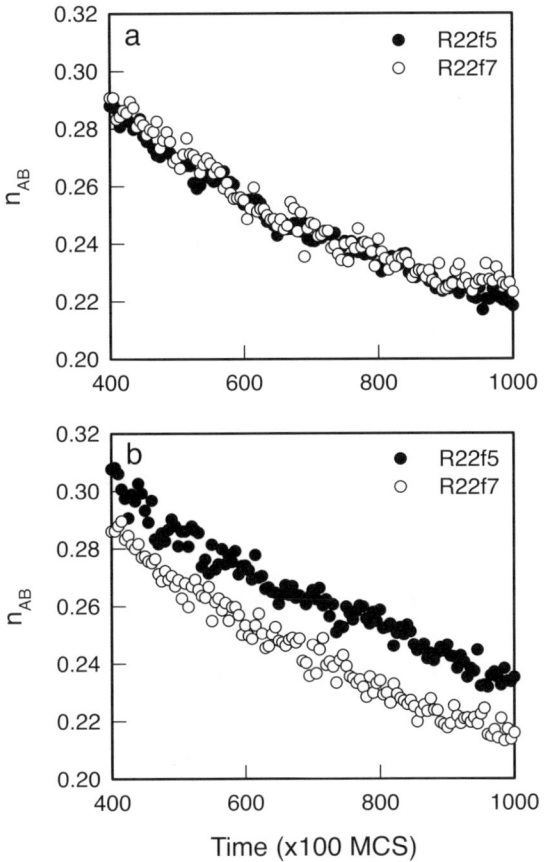

Fig. 22a,b. Change of normalized contact number n_{AB} during phase separation when the repulsive block copolymers of: **a** $N_{block} = 12$; **b** $N_{block} = 30$ with different composition f are added to A/B immiscible polymer blends [71]

time interval in which the complete universality of $F(x)$ is observed. In this stage, the time evolution of the self-assembled structure is scaled with a single length parameter over the entire length scale.

Figure 24 shows the time change of the scaled structure factor $F(x)$ for a ternary blend with a repulsive block copolymer (R22f5) with $N_{block} = 12$. $F(x)$ increases with time t up to 50,000 MC steps as shown in Fig. 24a, i.e., $F(x)$ is not universal. This results from the increase of the concentration fluctuation in the polymer mixture, and thus this stage corresponds to the intermediate stage of spinodal decomposition. With increasing time, $F(x)$ becomes universal with t as shown in Fig. 24b, implying that the dynamical self-similarity holds for this model system. This time regime corresponds to the late stage. The same scaling feature is also observed for other ternary blends containing the repulsive block

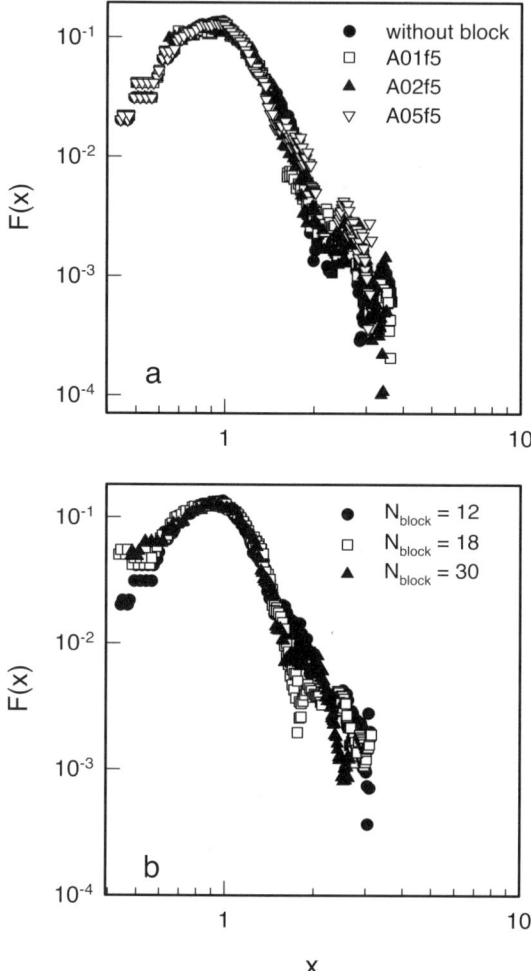

Fig. 23a,b. Scaled structure factor $F(x)$ as a function of the reduced scattering vector for A/B/C-b-D blend containing attractive block copolymers: **a** with different interaction energies and $N_{block} = 12$; **b** with different chain length of A05f5 block copolymers [68]

copolymers with different chain length and composition. The scaled structure factors are plotted against the reduced scattering vector in the double-logarithmic scale for various blend systems in Fig. 25. As shown in the figure, the scaled structure factors are universal at $x < 2$, irrespective of the chain length and the interaction type, indicating that the growth of phase-separated domains occurs with self-similarity for both cases of the attractive and repulsive block copolymers. In other words, the addition of block copolymers to immiscible blends does not affect the growth mechanism of global structure. These simulation re-

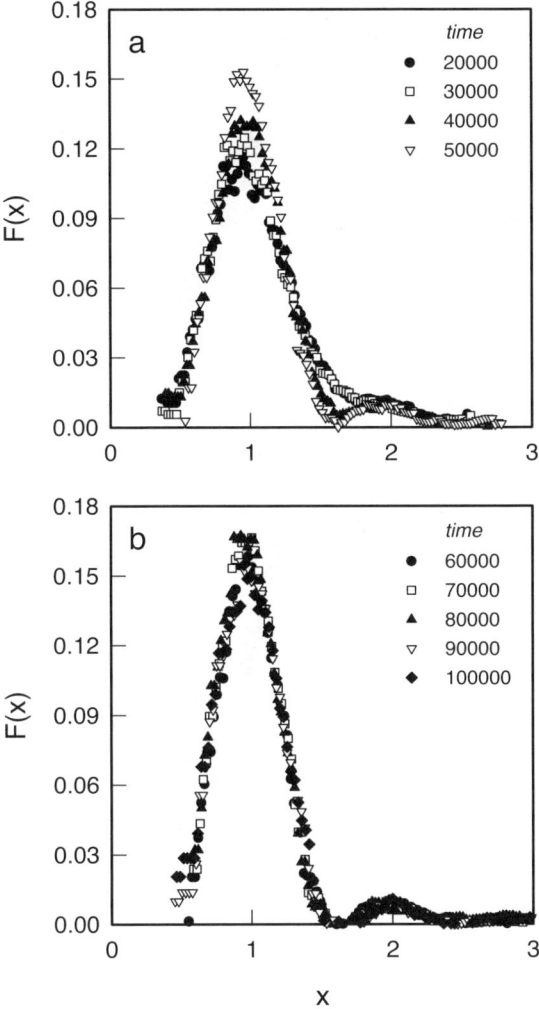

Fig. 24a,b. Scaled structure factor $F(x)$ as a function of the reduced scattering vector for A/B/C-b-D blend containing R22f5 block copolymer with $N_{block} = 12$ in: **a** the intermediate stage; **b** the late stage of phase separation [71]

sults are consistent with the experimental ones [58]. It is interesting to note that we can observe in Fig. 25 the peak around $x = 2$ which seems closely related to a locally patterned structure. This behavior can also be observed in the majority of binary mixtures without block copolymer [53–56].

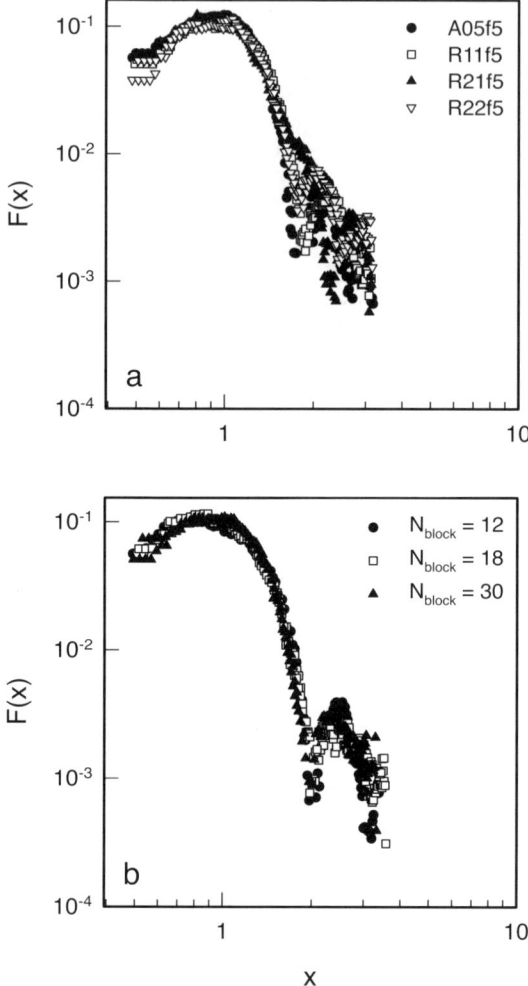

Fig. 25a,b. Scaled structure factor $F(x)$ as a function of the reduced scattering vector for A/B/C-b-D blend containing block copolymers: **a** with different interaction energies and $N_{block} = 30$; **b** with different chain length of R11f5 block copolymers [71]

4
Mechanical Properties of Semicrystalline Polymers

Molecular modeling techniques have been used to predict and interpret mechanical properties of polymers [88–95]. Theodorou and Suter [88, 89] found that the internal energy contribution to the elastic response is much more important than the entropic contribution for glassy polymers by a thermodynamic analysis. They modeled elastic properties of a glassy amorphous vinyl polymer using MM and found that atomic displacements deviate considerably from affine behavior. Fan and Hsu [93] predicted mechanical and thermal properties of the aromatic polysulfone relaxed by MD and MM. However, it has been very difficult, if not impossible, to predict mechanical properties of multiphase polymers such as polymer blends and semicrystalline polymers by molecular simulation due to its limited length scale ($\approx$ 30 Å).

There have been many efforts for combining the atomistic and continuum levels, as mentioned in Sect. 1. Recently, Santos et al. [11] proposed an atomistic-continuum model. In this model, the three-dimensional system is composed of a matrix, described as a continuum and an inclusion, embedded in the continuum, where the inclusion is described by an atomistic model. The model is validated for homogeneous materials (an fcc argon crystal and an amorphous polymer). Yang et al. [96] have applied the atomistic-continuum model to the plastic deformation of Bisphenol-A polycarbonate where an inclusion deforms plastically in an elastic medium under uniaxial extension and pure shear. Here the atomistic-continuum model is validated for a heterogeneous material and elastic constant of semicrystalline poly(trimethylene terephthalate) (PTT) is predicted.

4.1
Description of the Atomistic-Continuum Model

The system in the atomistic-continuum model is composed of a matrix, described as a continuum, and an inclusion represented in atomistic detail, as shown in Fig. 26. The matrix is modeled by the finite element method developed by Gusev [97]. The scaling matrix **H** = [**ABC**] describes the system under periodic boundary conditions, where **A**, **B**, and **C** are cell vectors [98–100]. A set of nod-

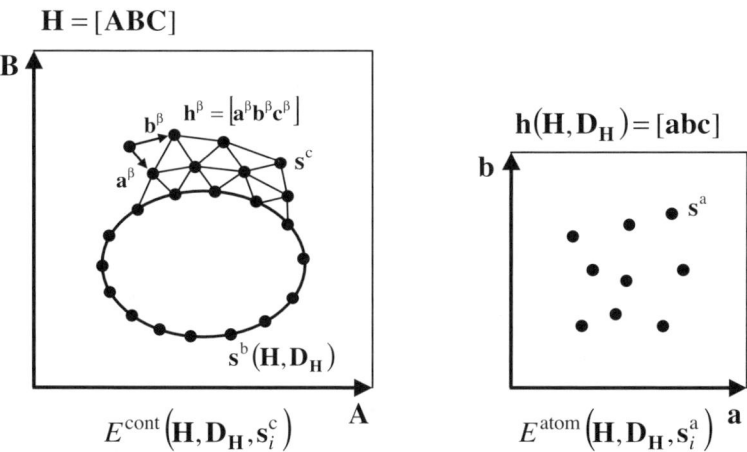

Fig. 26. Schematic diagram of the atomistic-continuum model. Note that s^b and h are dependent variable calculated from **H** and D_H

al points $\mathbf{x}_i^b$ and $\mathbf{x}_i^c$ are specified on the inclusion boundary and in the continuum, respectively, which are vertices of a periodic network of Delaunay tetrahedra [101]. The scaled coordinates $\mathbf{s}_i^b$ and $\mathbf{s}_i^c$ are used as degrees of freedom via $\mathbf{x}_i^b = \mathbf{H}\mathbf{s}_i^b$ and $\mathbf{x}_i^c = \mathbf{H}\mathbf{s}_i^c$. The local strain tensor inside each tetrahedron ε^β can be defined from the local scaling matrix $\mathbf{h}^\beta$ as

$$\varepsilon^\beta = \frac{1}{2}\left[\left(\mathbf{G}^\beta\right)^T \mathbf{G}^\beta - \mathbf{I}\right] \cdot \mathbf{G}^\beta = \mathbf{h}^\beta\left(\mathbf{h}_0^\beta\right)^{-1} \qquad (28)$$

where subscript 0 represents the undeformed state. The Helmholtz elastic free energy in the matrix is given by

$$E^{\text{cont}}\left(\mathbf{H}, \mathbf{s}_i^b, \mathbf{s}_i^c\right) = \frac{1}{2}\sum_\beta V_\beta \left(\mathbf{e}^\beta\right)^T \mathbf{C}^\beta \mathbf{e}^\beta \qquad (29)$$

where $\mathbf{e}^\beta$ is a six-dimensional strain vector obtained from $\mathbf{e}^\beta$. In atomistic simulation, the scaled coordinates of atoms are chosen as a degree of freedom via $\mathbf{x}_i^a = \mathbf{h}\mathbf{s}_i^a$, and therefore the internal energy in the atomistic box has the form of $E^{\text{atom}}(\mathbf{h}, \mathbf{s}_i^a)$.

The key idea to mix two different length scales is to couple the displacement of nodal points on the inclusion boundary with the change of the atomistic scaling matrix via

$$\mathbf{x}_i^a = \mathbf{h}\mathbf{h}_0^{-1}\mathbf{x}_{0,i}^b \qquad (30)$$

In other words, the inclusion boundary follows the homogeneous deformation of the atomistic box. Both the system box $\mathbf{H}$ and atomistic box $\mathbf{h}$ should be independent degrees of freedom in the model. Instead of considering $\mathbf{h}$ as a degree of freedom, the degree of freedom $\mathbf{D_H}$ is introduced to relate two scaling matrices through

$$\mathbf{h} = \mathbf{H}(\mathbf{I} + \mathbf{D_H})\mathbf{H}_0^{-1}\mathbf{h}_0 \qquad (31)$$

where $\mathbf{D_H}$ acts as the deviatoric part from $\mathbf{H}$. The dependent variable $\mathbf{s}_i^b$ is written as

$$\mathbf{s}_i^b = \mathbf{H}^{-1}\mathbf{x}_i^b = \mathbf{H}^{-1}\mathbf{h}\mathbf{h}_0^{-1}\mathbf{H}_0\mathbf{s}_{0,i}^b = (\mathbf{I} + \mathbf{D_H})\mathbf{s}_{0,i}^b \qquad (32)$$

Consequently, the system can be described by four types of degrees of freedom: $\mathbf{H}$, $\mathbf{h}$, $\mathbf{s}_i^c$, and $\mathbf{s}_i^a$. The total energy of the system is

$$E^{\text{sys}}\left(\mathbf{H}, \mathbf{D_H}, \mathbf{s}_i^c, \mathbf{s}_i^a\right) = E^{\text{cont}}\left(\mathbf{H}, \mathbf{D_H}, \mathbf{s}_i^c\right) + \alpha_V E^{\text{atom}}\left(\mathbf{H}, \mathbf{D_H}, \mathbf{s}_i^a\right) \quad \alpha_V = \frac{V^{\text{inc}}}{V^{\text{atom}}} \qquad (33)$$

where V^{inc} and V^{atom} are the volumes of inclusion and atomistic box, respectively. Note that α_V remains constant during the deformation because the inclusion boundary deforms according to the atomistic box. A detailed description of the model can be found in [11].

4.2
Simulation Methods

The system (semicrystalline PTT) is a cube with an edge length of 100 Å, which has one spherical inclusion (crystalline phase) embedded in the matrix (amorphous phase). The volume fraction of an inclusion is systematically varied to account for different crystallinity (f^{inc} = 0.11, 0.18, 0.27, and 0.38). In all cases, approximately 1500 nodal points are specified on the inclusion boundary ($\mathbf{x}_i^b$) and 6000 ones in the continuum ($\mathbf{x}_i^c$), resulting in approximately 45,000 tetrahedra after Delaunay tessellation. The Polak-Ribiere conjugate gradient algorithm [102] is adopted to minimize the system energy. Full atomistic model for amorphous PTT with 100 repeating units (2502 atoms) is prepared so as to yield the experimental density of 1.31 g cm^{-3}. Full atomistic model for crystalline PTT with 2400 atoms is also built by accumulating 48 unit cells from the experimentally determined scaled coordinates of atoms and unit cell parameters [103, 104]. Both atomistic boxes are equilibrated through MM and MD runs. The pcff force field [105] is used to evaluate the atomistic energy. All the results are reported by averaging three independent microstructures.

The constant strain method [11, 96] is used to calculate elastic constants. In this method, 21 independent sets of small strains (2.0×10^{-3}) are applied to the simulation box at the minimum energy configuration and the energy is minimized with respect to all remaining degrees of freedom at fixed shape of the simulation box (**h** for atomistic modeling, **H** for atomistic-continuum model). The elastic constants **C** are evaluated from the minimized energy E by

$$E - E_0 = \frac{1}{2} V \mathbf{e}^T \mathbf{C} \mathbf{e} \tag{34}$$

where V is the volume of the simulation box ($\|\mathbf{h}\|$ for atomistic modeling, $\|\mathbf{H}\|$ for atomistic-continuum model). For convenience of calculation, the stiffness matrix C_{ij} is converted into the compliance matrix S_{ij} from which Young's modulus is calculated by

$$E_1 = S_{11}^{-1}, \; E_2 = S_{22}^{-1}, \; E_3 = S_{33}^{-1} \tag{35}$$

The whole simulations are performed according to the following procedure. First, elastic constants of PTT in amorphous phase $\mathbf{C}^{amor}$ are calculated using the atomistic modeling, which will be used as input values for the matrix $\mathbf{C}^\beta$ in the atomistic-continuum model. Second, elastic constants of PTT in crystalline phase $\mathbf{C}^{crst}$ are also evaluated in the same manner as those of amorphous PTT. Third, the atomistic-continuum model is validated for heterogeneous material by comparing the calculated elastic constant for the system of infinite lamellas with its exact solution. Finally, elastic constants of semicrystalline PTT with different crystallinity can be evaluated using the atomistic-continuum model. The crystallinity is controlled by varying the volume fraction of the inclusion.

4.3
Results and Discussion

The stiffness matrices of amorphous and crystalline PTT calculated by atomistic modeling are

$$C_{ij}^{amor} = \begin{pmatrix} 9.65 & 4.65 & 4.80 & -0.16 & 0.33 & 0.46 \\ 4.65 & 9.61 & 5.10 & -0.45 & -0.03 & 0.30 \\ 4.80 & 5.10 & 9.08 & -0.46 & -0.25 & 0.32 \\ -0.16 & -0.45 & -0.46 & 2.51 & 0.18 & 0.16 \\ 0.33 & -0.03 & -0.25 & 0.18 & 2.19 & -0.20 \\ 0.46 & 0.30 & 0.32 & 0.16 & -0.20 & 2.42 \end{pmatrix} (GPa) \text{ for amorphous PTT} \quad (36)$$

$$C_{ij}^{amor} = \begin{pmatrix} 25.78 & 13.38 & 16.89 & 0.47 & -1.89 & 2.67 \\ 13.38 & 24.85 & 16.79 & 2.60 & 0.50 & 1.34 \\ 16.89 & 16.79 & 35.11 & -1.24 & -1.84 & -0.46 \\ 0.47 & 2.60 & -1.24 & 8.26 & -2.43 & 0.34 \\ -1.89 & 0.50 & -1.84 & -2.43 & 10.44 & -0.34 \\ 2.67 & 1.34 & -0.46 & 0.34 & -0.34 & 4.94 \end{pmatrix} (GPa) \text{ for crystalline PTT} \quad (37)$$

For the system of infinite lamellas, the exact solution for C_{33} is obtained from Eq. (38) [97]:

$$\frac{1}{C_{33}^{exact}} = \frac{f^{inc}}{C_{33}^{amor}} + \frac{(1-f^{inc})}{C_{33}^{crst}} \quad (38)$$

where f^{inc} is the volume fraction of lamellar inclusion. If the system is deformed only in the 3-direction, the exact value of strain energy E^{exact} will be

$$E^{exact} = \frac{1}{2} V C_{33}^{exact} e_3^2 \quad (39)$$

Figure 27 shows that the simulated strain energy converges to its exact value obtained from Eqs. (38) and (39) as the energy minimization proceeds. Thus it is clear that the atomistic-continuum model has the consistency for heterogeneous materials as well as for homogeneous materials.

If the system undergoes deformation in the atomistic-continuum model, all variables are initially set to deform affinely according to **H**, and then can vary from their initial positions during energy minimization. Therefore, the initial

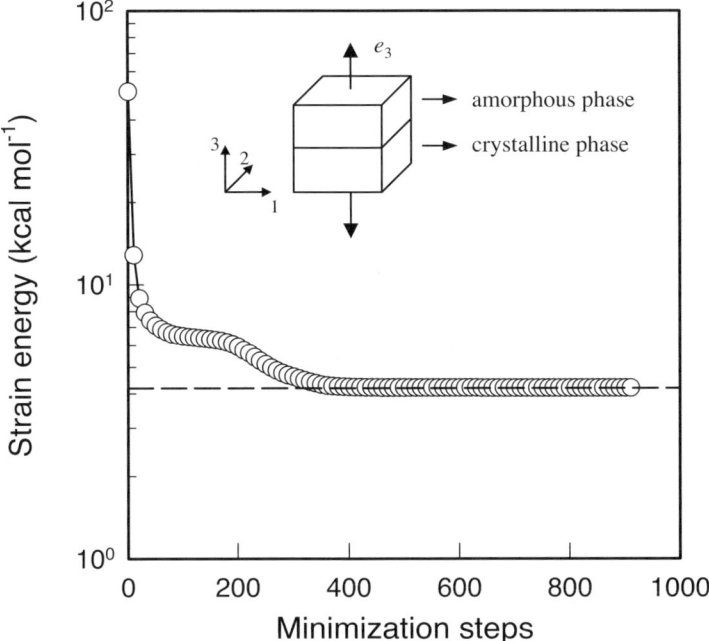

Fig. 27. Validation of the atomistic-continuum model for the system of infinite lamellas of f^{inc}. The *dashed line* represents the exact value of the strain energy obtained from Eqs. (38) and (39)

value of the inclusion energy, which is obtained from the atomistic energy, is much higher than that of the matrix energy due to the connectivity of atoms, as shown in Fig. 28. Another interesting feature from Fig. 28 is that the inclusion energy decreases while the matrix energy increases continuously during energy minimization. This is because the inclusion (crystalline phase) is much stiffer than the matrix (amorphous phase), as clearly seen in Eqs. (36) and (37). The strain of the atomistic box in the deformed direction decreases from its initial value, as shown in Fig. 29, resulting in the decrease in the inclusion energy. Because the system strain does not vary during energy minimization, the decrease in the strain of atomistic box leads to an increase in the strain of tetrahedra in the continuum, which subsequently results in the increase in the matrix energy. Figure 30 shows that the modulus in the 3-direction of semicrystalline PTT increases with crystallinity, as expected. The simulated modulus lies in between the Voigt model (parallel model) and the Reuss model (series model). It is important to define a representative volume element (RVE) which is large enough to represent the overall behavior of heterogeneous materials [97, 106]. Gusev [97] compared the overall elastic constants of the systems with 1, 8, 27, and 64 spherical inclusions and found that the RVE size is remarkably small. However, in his study, elastic constants calculated from the system with one sphere

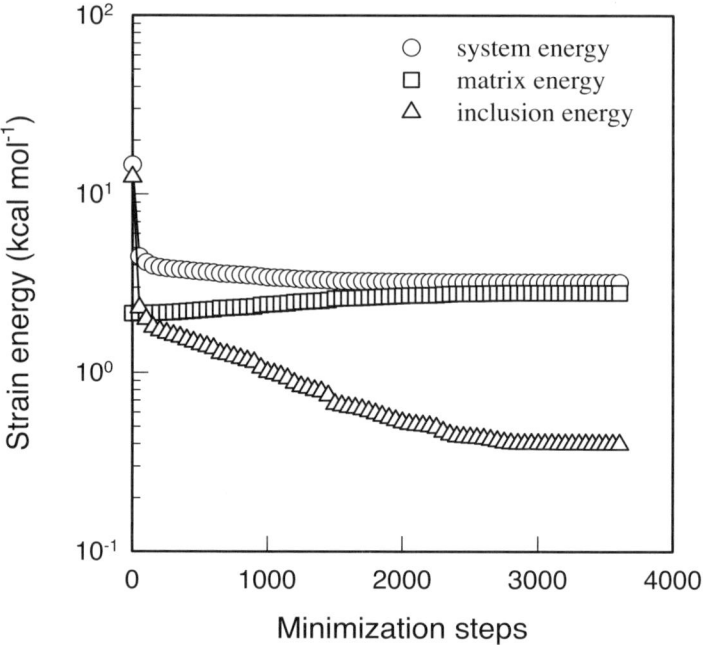

Fig. 28. Change of strain energy components during energy minimization under deformation in the 3-direction ($e_3^{sys} = 2.0 \times 10^{-3}$) for the system containing one spherical inclusion ($f^{inc} = 0.18$)

showed deviation from the averages by 10%. If the system including more than one spheres is used in this simulation, more precise value can be obtained.

5
Conclusions

This review has illustrated various properties of multiphase polymer systems obtained from computer simulation. Three modeling techniques – atomistic, coarse-grained, and atomistic-continuum modeling – are applied to miscibility of homopolymer/copolymer and homopolymer/homopolymer blends, compatibilizing effect of block copolymers, and mechanical properties of semicrystalline polymers, respectively.

As an example of atomistic modeling for multiphase polymer systems, miscibility of PEO/SAA and PS/PVME blends are investigated. For PEO/SAA blends, the effect of sequence distribution of copolymer on the miscibility of blends is analyzed by calculating the interaction energy parameters. It is observed that both the sequence distribution and the composition significantly affect the degree of miscibility. For a fixed composition, there exists an optimal range of sequence distribution for which the blend system is miscible. The sequence distri-

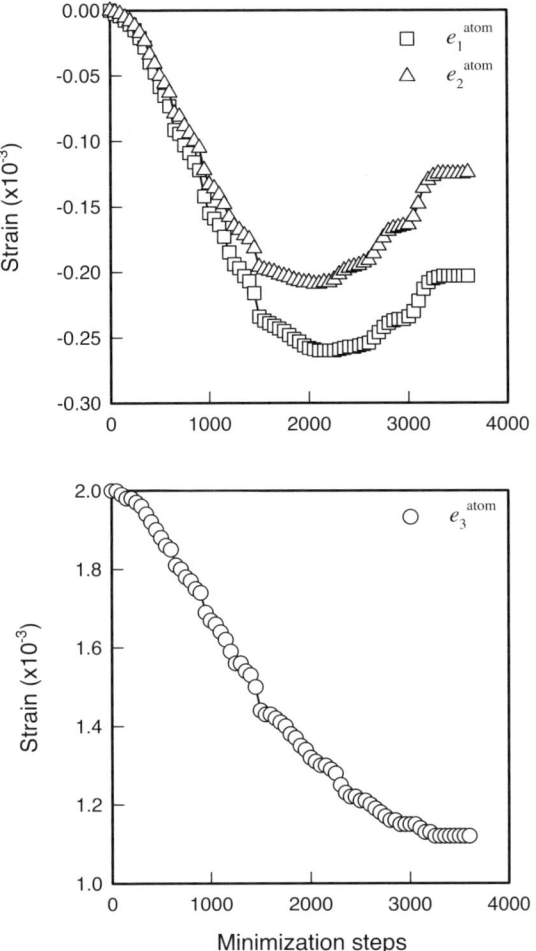

Fig. 29. Change of strains of the atomistic box during energy minimization under deformation in the 3-direction ($e_3^{sys} = 2.0 \times 10^{-3}$) for the system containing one spherical inclusion ($f^{inc} = 0.18$). Note that the system strains are fixed at their initial values in the constant strain method

bution not only affects the charge distribution of segments which in turn affects the contact energy, but also affects the probability of contacts between interaction sites. It is also observed that the segmental interaction energy itself is more important than the local spatial distribution of the segments, to determine the effect of the copolymer sequence distribution on the miscibility.

For PS/PVME blends, the thermodynamic properties are calculated by MD and MM, from which the characteristic parameters of the equation-of-state theory, p^*, v_{sp}^*, and T^* are determined based on the physical meaning of the para-

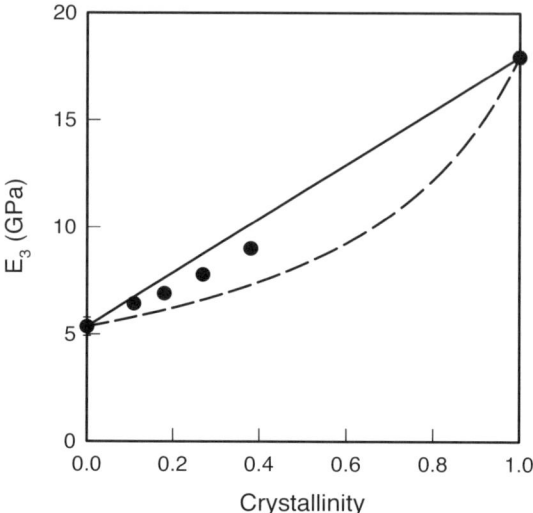

Fig. 30. Young's modulus in the 3-direction E_3 for semicrystalline PTTs with different crystallinity. The *solid and dashed lines* represent the Voigt and Reuss averages, respectively [106]

meters. The lattice fluid theory with simulated characteristic parameters is used for the prediction of the surface tension of each polymer and the phase diagram of the blend. The calculated surface tensions of PS and PVME with no adjustable parameter agree well with the experimental data within ca. 1.0 mN m^{-1}, indicating that the simulated equation-of-state parameters for the pure polymers are reasonable. The calculated phase diagram of PS/PVME blends is also comparable to the experimental one with the use of an adjustable parameter ζ.

As a typical example of coarse-grained modeling for multiphase polymer systems, the phase separation dynamics of immiscible polymer blends in the presence of block copolymers as a compatibilizer is investigated via MC simulations. The block copolymers used for simulation are composed of chemically different blocks (C and D) from homopolymers (A and B). For comparison, the phase separation dynamics of A/B binary mixture is also investigated. For A/B binary blends, the typical behavior of spinodal decomposition is observed in simulations, and three stages of spinodal decomposition, i.e., early, intermediate, and late stages are clearly identified. In the late stage, the binary system obeys the dynamical scaling law, which is consistent with the experimental results.

For A/B/C-*b*-D ternary blends, it is generally found that the block copolymer can significantly suppress the growth rate of phase-separated structure, due to the reduction of interfacial tension. In particular, even a repulsive block copolymer can considerably retard the phase separation process if the interaction energies in the system satisfy a proper condition. However, such a retardation effect by a block copolymer is highly dependent on the molecular structure of block copolymer, such as its chain length and composition. The attractive and repulsive block copolymers show different behaviors in phase separation. For example, an attractive block copolymer with asymmetric composition ($f = 0.7$) shows a better retardation effect on phase separation of A/B (75/25) blend than the symmetric one ($f = 0.5$), whereas a repulsive block copolymer with the asymmetric composition does not show a favorable retardation effect as compared with the symmetric one. As the case of A/B binary blend, the dynamical scaling law was applied to ternary blends containing both attractive and repulsive block copolymers in the late stage of phase separation. It is observed that the scaled structure factors obtained from all blend systems are universal with time, irrespective of the chain length and composition of block copolymer, and its interaction type with homopolymers. In other words, there exists self-similarity between the structures developed at different time scales in different systems, implying that the addition of block copolymer to immiscible blends does not affect the growth mechanism of phase-separated structures.

As an example of atomistic-continuum modeling for multiphase polymer systems, elastic constant of semicrystalline PTT is calculated and compared with its analytical solution. As the energy minimization proceeds, perfect agreement between simulated strain energy and its exact solution is observed. The system for semicrystalline PTT consists of a matrix (amorphous phase), described as a continuum, and a spherical inclusion (crystalline phase) represented in atomistic detail. The crystallinity of PTT can be controlled by varying the volume fraction of an inclusion. The Young's modulus of semicrystalline PTT increases with crystallinity, as expected. The system studied is thought to be smaller than the RVE size. However, it is straightforward to simulate the system including several spheres in the atomistic-continuum model, although it needs considerable computing time.

References

1. Allen MP, Tildesley DJ (1987) Computer simulation of liquids. Clarendon, Oxford
2. Monnerie L, Suter UW (ed) (1994) Advances in polymer science, vol 116. Springer, Berlin Heidelberg New York
3. Binder K (1995) Monte Carlo and molecular dynamics simulations in polymer science. Oxford, Oxford
4. Leach AR (1996) Molecular modelling. Longman, Harlow
5. Frenkel D, Smit B (1996) Understanding molecular simulation. Academic Press, San Diego
6. Binder K, Heermann DW (1988) Monte Carlo simulation in statistical physics. Springer, Berlin
7. Tadmor EB, Phillips R, Ortiz M (1996) Langmuir 12:4529
8. Tadmor EB, Ortiz M, Phillips R (1996) Phil Mag A 73:1529
9. Shenoy VB, Miller R, Tadmor EB, Phillips R, Ortiz M (1998) Phys Rev Lett 80:742
10. Rafii-Tabar H, Hua L, Cross M (1998) J Phys Condens Matter 10:2375
11. Santos S, Yang JS, Gusev AA, Jo WH, Suter UW (2001) (submitted)
12. ten Brinke G, Karasz FE, Macknight WJ (1983) Macromolecules 16:1827
13. Paul DR, Barlow JW (1984) Polymer 25:487
14. Kambour RP, Bendler JT, Bopp RC (1983) Macromolecules 16:753

15. Jo WH, Lee SC (1990) Macromolecules 23:2261
16. Balazs AC, Sanchez IC, Epstein IR, Karasz FE, Macknight WJ (1985) Macromolecules 18:2188
17. Sanchez IC (1978) In: Paul DR (ed) Polymer blends. Academic Press, New York, chap 3
18. Sanchez IC, Lacombe RH (1976) J Phys Chem 80:2352
19. Sanchez IC, Lacombe RH (1978) Macromolecules 11:1145
20. Lacombe RH, Sanchez IC (1976) J Phys Chem 80:2568
21. Sanchez IC, Lacombe RH (1977) J Polym Sci Polym Lett Ed 15:71
22. Choi K, Jo WH (1997) Macromolecules 30:1509
23. Choi K, Jo WH, Hsu SL (1998) Macromolecules 31:1366
24. Poser CI, Sanchez IC (1979) J Colloid Interface Sci 69:539
25. Sanchez IC (1992) In: Sanchez IC (ed) Physics of polymer surfaces and interfaces. Butterworth-Heinemann, Stoneham, chap 5
26. Jo WH, Choi K (1997) Macromolecules 30:1800
27. Fan CF, Olafson BD, Blanco M, Hsu SL (1992) Macromolecules 25:3667
28. Tao H-J, Fan CF, Macknight WJ, Hsu SL (1994) Macromolecules 27:1720
29. Rappé AK, Goddard WA III (1991) J Phys Chem 95:3358
30. Boyd RH, Pant PVK (1991) Macromolecules 24:4078
31. Cantow H-J, Schulz O (1986) Polym Bull 15:449
32. Walsh DJ, Dee GT, Halary JL, Ubiche JM, Millequant M, Lesec J, Monnerie L (1989) Macromolecules 22:3395
33. Simha R, Somcynski T (1969) Macromolecules 2:342
34. Dee GT, Sauer BB (1993) Macromolecules 26:2771
35. Shiomi T, Hamada F, Nasako T, Yoneda K, Imai K, Nakajima A (1990) Macromolecules 23:2296
36. Garcia D (1984) J Polym Sci Polym Phys Ed 22:1733
37. Han CC, Baurer BJ, Clark JC, Muroga Y, Matsushita Y, Okada M, Tran-Cong Q, Chang T, Sanchez IC (1988) Polymer 29:2002
38. Guggenheim EA (1935) Proc R Soc London A148:304
39. Sanchez IC, Balazs AC (1989) Macromolecules 22:2325
40. de Gennes P-G (1980) J Chem Phys 72:4756
41. Pincus P (1981) J Chem Phys 75:1996
42. Binder K (1983) J Chem Phys 79:6387
43. Hashimoto T, Kumaki J, Kawai H (1983) Macromolecules 16:641
44. Synder HL, Meakin P, Kawai H (1983) Macromolecules 16:757
45. Gelles R, Frank CW (1983) Macromolecules 16:1448
46. Hashimoto T, Sasaki K, Kawai H (1984) Macromolecules 17:2812
47. Strobl GR (1985) Macromolecules 18:558
48. Izumitani T, Hashimoto T (1985) J Chem Phys 83:3694
49. Inoue T, Ougisawa T, Yasuda O, Miyasaka K (1985) Macromolecules 18:78
50. Hashimoto T, Itakura M, Hasegawa H (1986) J Chem Phys 85:6118
51. Hashimoto T, Itakura M, Shimidzu N (1986) J Chem Phys 85:6773
52. Nose T (1987) Phase Transitions 8:245
53. Bates FS, Wiltzius P (1989) J Chem Phys 91:3258
54. Chakrabarti A, Toral R, Gunton JD, Muthukumar M (1990) J Chem Phys 92:6899
55. Hashimoto T, Takenaka M, Jinnai H (1991) J Appl Cryst 24:457
56. Takenaka M, Hashimoto T (1992) J Chem Phys 96:6177
57. Nakai A, Shiwaku T, Wang W, Hasegawa H, Hashimoto T (1996) Macromolecules 29:5990
58. Roe RJ, Kuo CM (1990) Macromolecules 23:4635
59. Park DW, Roe RJ (1991) Macromolecules 24:5324
60. Hashimoto T, Izumitani (1993) Macromolecules 26:3631
61. Izumitani T, Hashimoto T (1994) Macromolecules 27:1744
62. Laradji M, Guo H, Grant M, Zuckermann M (1992) J Phys Condens Matter 4:6715

63. Kawakatsu T, Kawasaki K, Furusaka M, Okabayashi H, Kanaya T (1993) J Chem Phys 99:8200
64. Lin CC, Jeon HS, Balsara NP, Hammouda B (1995) J Chem Phys 103:1957
65. Jo WH, Kim HC, Baik DH (1991) Macromolecules 24:2231
66. Kim HC, Nam KH, Jo WH (1993) Polymer 34:4043
67. Jo WH, Jo BC, Cho JC (1994) J Polym Sci Polym Phys Ed 32:1661
68. Jo WH, Kim SH (1996) Macromolecules 29:7204
69. Kim SH, Jo WH, Kim J (1996) Macromolecules 29:6933
70. Kim SH, Jo WH, Kim J (1997) Macromolecules 30:3910
71. Kim SH, Jo WH (1998) J Chem Phys 108:4267
72. Carmesin I, Kremer K (1988) Macromolecules 21:2819
73. Deutsch HP, Binder K (1991) J Chem Phys 94:2294
74. Chan JW (1965) J Chem Phys 42:93
75. Cook HE (1970) Acta Metall 18:297
76. Marro J, Bortz AB, Karlos MH, J. Lebowitz JL (1975) Phys Rev B 12:2000
77. Binder K, Karlos MH, Lebowitz JL, Marro J (1979) Adv Colloid Interface Sci 10:173
78. Lebowitz JL, Marro J, Kalos MH (1982) Acta Metall 30: 297
79. Lifshitz IM, Slyozov VV (1961) J Phys Chem Solids 19:35
80. Koga T, Kawasaki K (1991) Phys Rev A 44:R817
81. Vilgis TA, Noolandi J (1990) Macromolecules 23:2941
82. Noolandi J, Hong KM (1982) Macromolecules 15:482
83. Noolandi J, Hong KM (1984) Macromolecules 17:1531
84. Vilgis TA, Noolandi J (1990) Macromolecules 23:2941
85. Israels R, Jasnow D, Balaz AC, Guo L, Krausch G, Sokolov J, Rafailovich M (1995) J Chem Phys 102:8149
86. Leibler L (1980) Macromolecules 13:1602
87. Tanaka H, Hashimoto T (1988) Polym Commun 29:212
88. Theodorou DN, Suter UW (1986) Macromolecules 19:139
89. Theodorou DN, Suter UW (1986) Macromolecules 19:379
90. Mott PH, Argon AS, Suter UW (1993) Phil Mag A 67:931
91. Hutnik M, Argon AS, Suter UW (1993) Macromolecules 26:1097
92. Brown D, Clarke JHR (1991) Macromolecules 24:2075
93. Fan CF, Hsu SL (1992) Macromolecules 25:266
94. Jang SS, Jo WH (1999) Macromol Theory Simul 8:1
95. Jang SS, Jo WH (1999) J Chem Phys 110:7524
96. Yang JS, Jo WH, Santos S, Suter UW (2001) In: Kotelyanskii M, Theodorou DN (ed) Simulation methods for modeling polymers. Marcel Dekker, (submitted)
97. Gusev AA (1997) J Mech Phys Solids 45:1449
98. Parrinello M, Ramen A (1980) Phys Rev Lett 45:1196
99. Parrinello M, Ramen A (1981) J Appl Phys 52:7182
100. Parrinello M, Ramen A (1982) J Chem Phys 76:2662
101. Tanemura M, Ogawa T, Ogita N (1983) J Comp Phys 51:191
102. Press WH, Teukolshy SA, Vetterling WT, Flannery BP (1992) Numerical recipes, 2nd edn. Cambridge University Press, Cambridge
103. Poulin-Dandurand S, Pérez S, Revol J-F, Brisse F (1979) Polymer 20:419
104. Desborouh IJ, Hall IH, Neisser JZ (1979) Polymer 20:545
105. Maple JR, Hwang M-J, Stockfisch TP, Dinur U, Waldman M, Ewig CS, Hagler AT (1994) J Comp Chem 15:162
106. Tsai SW, Hahn HT (1980) Introduction to composite materials. Technomic Publishing Company, Westport

Editor: Prof. K.-S. Lee
Received: February 2001

Adhesion and Fracture of Interfaces Between Immiscible Polymers: from the Molecular to the Continuum Scale

Costantino Creton[1], Edward J. Kramer[2], Hugh R. Brown[3], Chung-Yuen Hui[4]

[1]Laboratoire de Physico-Chimie Structurale et Macromoléculaire, ESPCI, 10 Rue Vauquelin, 75231 Paris Cédex 05, France
[2]University of California, Santa Barbara, Department of Materials Engineering II, 1361 C, Santa Barbara, CA 93106, USA
[3]BHP Steel Institute, University of Wollongong, Wollongong, NSW 2522, Australia
[4]Department of Theoretical and Applied Mechanics, Cornell University, Ithaca, NY 14853, USA
e-mail: [1]costantino.creton@espci.fr, [2]Kramer@msc.cornell.edu

In order to obtain a measurable fracture toughness, a joint between two immiscible polymer glasses must be able to transfer mechanical stress across the interface. This stress transfer capability is very weak for narrow interfaces and a significant reinforcement can be achieved, either by the use of connecting chains (block copolymers), or by a broadening of the interface (random copolymers). In both cases, the stress is transferred by entanglements between polymer chains. The molecular criteria for efficient stress transfer, by connecting chains and by broad interfaces, are reviewed here with a special emphasis on the role of the molecular architecture (diblock, triblock or random copolymers) and molecular weight of the chains present at the interface. Recent theoretical developments in the relationship between macroscopic fracture toughness and interfacial stress transfer are also discussed, and the essential role of bulk plastic deformation properties of the polymers on either side of the interface are specifically addressed.

Keywords. Fracture, Polymer interfaces, Adhesion, Crazing

1	Introduction	56
2	Experimental Methods	60
2.1	Synthetic Methods and Interface Formation	61
2.2	Surface Analysis and Characterization	63
2.3	Mechanical Testing of the Interface and Fracture Mechanics.	64
3	Simple Connecting Chains Between Glassy Polymers	68
3.1	Main Experimental Results.	69
3.2	Fracture Mechanisms: Chain Pullout	72
3.3	Fracture Mechanisms: the Transition from Simple Chain Pullout to Failure by Crazing	77
3.4	Fracture Mechanisms: the Transition from Simple Chain Scission to Failure by Crazing	79
3.5	Fracture Mechanisms: Craze Growth and Stability	83
3.6	Effect of the Homopolymer.	92
3.7	Velocity Dependence	100

3.8	Effect of an Elastomeric Midblock Within the Connecting Chains	101
4	**Optimum Toughening of the Interface: the Limits of G_c**	**103**
4.1	Diblock Copolymers	105
4.2	A-B-A Triblock Copolymers	108
5	**Interfaces Between Polymers, Coupling by Random Copolymers**	**111**
5.1	Interfaces Between Homopolymers	111
5.2	Random Copolymers at Interfaces Between Homopolymers	115
6	**Reactive Systems**	**121**
6.1	Reactive Polymers with Multiple Functional Groups	124
6.2	Interfaces Between Semicrystalline Polymers	125
7	**Conclusions**	**129**
References		**133**

Abbreviations

a	crack length
$\dot{a}$	crack velocity
a_I	interface width between two polymers
b	statistical segment length
C_{22}, C_{66}	longitudinal modulus (along main fibrils), shear modulus of a craze
d	distance between main craze fibrils
d_e	rms end-to-end distance between entanglements
E_i	Young's modulus of polymer i
f	force on a connector chain at interface
f	mole fraction of a mer in a random copolymer
f_b	force to break a single C–C bond in the polymer backbone
f_{mono}	friction force per monomer on a connector chain being pulled out
ΔG	free energy of mixing of two polymers
G	strain energy release rate of crack at interface
G_c	critical strain energy release rate or fracture toughness of interface
G_c^*	standard fracture toughness corresponding to the standard crack tip opening displacement
h	width of the craze normal to the interface at crack tip
h_i	thickness of beam i of ADCB sample
k_B	Boltzmann's constant
K	complex stress intensity factor
K_1	real part of complex stress intensity factor (controls normal stress)

K_2	imaginary part of complex stress intensity factor (controls shear stress)
K_{tip}	stress intensity factor at the crack tip within the continuum model of a craze
M_e	entanglement molecular weight
n	number of main chain bonds between crosslinks or entanglements
N	degree of polymerization of a polymer or polymer block
N_e	entanglement degree of polymerization of polymer or polymer block (corresponds to M_e)
T	temperature in Kelvin
U	energy to break a main chain bond
δ_c	continuum opening displacement at the crack tip with a craze
$\delta\ddagger$	standard crack tip opening displacement
Δ	wedge thickness for ADCB test
ε	dimensionless material constant given by Eq. (4)
κ_i	dimensionless elastic constant of polymer i ($= 3 - 4\nu_i$ for plane stress and $= (3 - \nu_i)/(1 + \nu_i)$ for plane strain)
μ_i	shear modulus of polymer i
$\sigma_{craze} = \sigma_c$	stress for craze widening
Σ	areal chain density of block copolymer or end-grafted polymer chains
Σ^*	areal chain density at transition from simple chain scission to crazing at the interface
$\Sigma\dagger$	areal chain density at transition from chain pullout to crazing at the interface
Σ_{sat}	saturation areal chain density of block copolymer beyond which more chains cannot be accommodated at the interface
Σ_{eff}	effective number of connector strands per unit area within the craze at the crack tip
Σ_{max}	maximum attainable areal chain density of connectors
σ_{fibril}	failure stress of the craze fibril structure at the crack tip
ν_i	Poisson's ratio of material i
υ_f	volume fraction of fibrils in craze
φ	volume fraction of polymer
χ	Flory segment-segment interaction parameter
ψ	mechanical phase angle (often called mode mixity)
ADCB=DCB	asymmetric double cantilever beam fracture sample or test
dPS	poly(d8-styrene) = deuterated polystyrene
FRES	forward recoil spectrometry
HIPS	high impact (rubber modified) poly(styrene)
MPS	monodisperse poly(styrene)
NRA	nuclear reaction analysis
PA-6	polyamide 6
PB	poly(butadiene)
PEB	poly(ethylene-r-butylene) random copolymer

PMMA	poly(methylmethacrylate)
PP	poly(propylene)
PP-g-MA	poly(propylene graft maleic anhydride)
PPO	poly(2,6-dimethyl-1,4-phenylene oxide)
PSAN	poly(styrene-r-acrylonitrile)
PS-COOH	PS chain with a carboxylic acid end-group
PS	poly(styrene)
PS(OH)	poly(styrene-r-4-hydroxystyrene)
PVP	poly(2-vinylpyridine)
SMA	poly(styrene-r-maleic anhydride)
SIMS	secondary ion mass spectrometry
TEM	transmission electron microscopy
XEp	crosslinked epoxy network
XPS	X-ray photoelectron spectrometry

1
Introduction

The adhesion between different polymers as well as between polymers and non-polymers is important for a wide range of applications. Most adhesives are polymeric in character. In two-phase polymer alloys, such as the rubber-toughened polymers in widespread use as engineering thermoplastics, adhesion at the phase boundaries is an important issue. In electronics the increasing use of polymers as dielectrics in multilayer structures means that adhesion at the interfaces thus created is of strong concern, especially because such interfaces must often survive accelerated testing involving thermal and humidity cycling to ensure their reliability. In the rubber industry the adhesion of different layers of rubber during building of tires and of rubber to tire cord is important. However, strong adhesion is not always desired and, in fact, non-stick coatings are usually also made of polymers.

In all of these applications, adhesion is caused by molecular interactions at the interface. Generally, these interactions cannot be probed directly since the formation of the interface is not a reversible process in the thermodynamic sense. Adhesion is therefore evaluated quantitatively with a destructive test of the interface. Ideally, one would like to be able to relate this measured adhesion to the underlying chemistry and physics of the interface itself. This fundamental approach, while it has been confined, for the most part, to interfaces between model polymers, has increasingly shed light on the molecular requirements for adhesion. At the same time, rapid advances in understanding of the micromechanics of cracking at interfaces between dissimilar materials has paved the way for theoretical approaches and simulation techniques that can bridge from the macroscopic descriptions of continuum solid mechanics to molecular lengths scales at the crack tip. Finally, the microstructural aspects of non-elastic deformation at the crack tip in polymers have gradually become clear, especially for glassy polymers, through the use of transmission electron microscopy and X-ray scattering techniques. These developments have laid the groundwork for the experiments and theory we will describe in this review.

Since we will focus below primarily on adhesion at interfaces between two immiscible polymers, it is appropriate to describe briefly what is known about such interfaces. The Gibbs free energy of mixing (per segment) of any two homopolymers A and B is given approximately by the Flory-Huggins expression:

$$\frac{\Delta G(\varphi)}{k_B T} = \frac{\varphi}{N_A}\ln\varphi + \frac{1-\varphi}{N_B}\ln(1-\varphi) + \chi\varphi(1-\varphi) \tag{1}$$

where N_A and N_B are the degree of polymerization of polymers A and B, φ is the volume fraction of polymer A segments and χ is the Flory segment-segment interaction parameter. Since χ between any two polymers chosen at random is usually positive, Eq. (1) implies a strong immiscibility at typical polymer lengths $N_A \approx N_B = N \sim 1000$. In the limit $\chi N \gg 1$, the segment volume fraction profile along a coordinate z normal to the interface is given by [1]:

$$\varphi(z) = \frac{1}{2}\left(1 - \tanh\left(\frac{2z}{a_I}\right)\right) \tag{2}$$

where a_I is an interface width given by:

$$a_I = \frac{2b}{\sqrt{6\chi}} \tag{3}$$

and where b is an appropriately averaged statistical segment length of the two polymers. Figure 1 shows the segment volume fraction profile calculated for a pair of immiscible polymers, polystyrene (PS) and poly(2-vinylpyridine) (PVP), that will be one of the model interfaces discussed in detail below. Note that the interface in this case is less than 2 nm wide.

The very small widths of such interfaces lead to little penetration of A chains into B chains, and vice versa, and thus very few entanglements are made across the interface. This lack of entanglement across the interface is thought to be responsible for the very low adhesion, as represented by the fracture energy G_c of such interfaces. For example, the G_c of the PS/PVP interface is about 1.5 J/m^2 whereas the G_c of the PS homopolymer is ~500–1000 J/m^2. Significantly, a polymer pair with a smaller positive value of χ, PS/poly(methylmethacrylate) (PMMA) which has a $\chi \sim 0.03$ and thus a value of $a_I \sim 3.2$ nm, has a considerably larger $G_c \sim 10$ J/m^2.

Thus, increasing the adhesion of such interfaces generally requires us to replace the entanglement network strands (or crosslinked strands in a crosslinked polymer) that would naturally span the plane of the interface if the same polymer were on both sides with some sort of molecular connections. These can be produced by block copolymers that reside at the interface and entangle with the polymers on either side, or can be produced by chemical reaction at the inter-

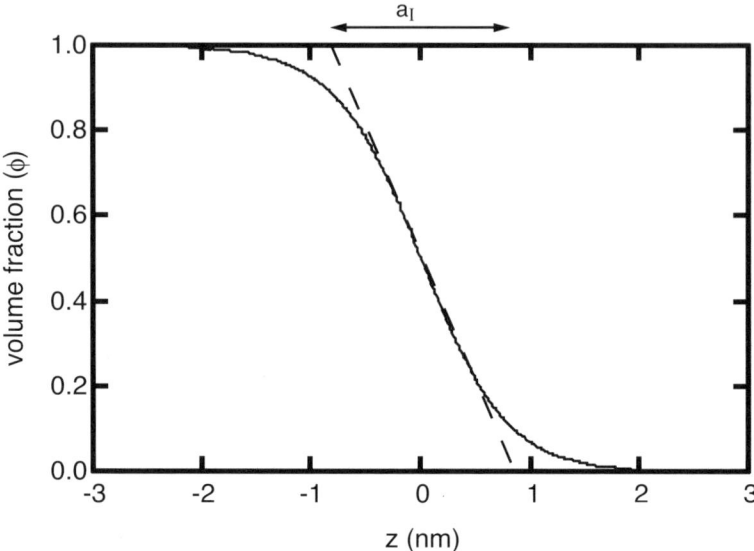

Fig. 1. Theoretical volume fraction vs. distance profile across an interface between PS and PVP polymers. A value of $\chi = 0.12$ appropriate to a temperature of 160 °C [105] has been used to predict an interface width a_I of 1.6 nm from Eq. (3). This profile is consistent with neutron reflectivity measurements of PS/PVP interface segment density profiles if the apparent broadening of the interface by capillary waves is taken into account [106]

face. For smooth polymer/nonpolymer interfaces where no polymer penetration into the nonpolymer side is possible, the formation of chemical bonds due to specific interactions of an acid-base or hydrogen-bonding nature between the polymer and the nonpolymer may be required in the absence of covalent bonding across the interface. As we will show below, however, the details of molecular connection at the interface are only part of the story. The non-elastic deformation of the polymer on at least one side of the interface is important for achieving substantial adhesion, as is an understanding of the mechanics of fracture of these interfaces.

The adhesion between immiscible, glassy polymers has only been investigated quantitatively relatively recently. One of the recurring problems has been to find a testing method that would give results that are characteristic of the interface and not of the experimental geometry. Recent progress in describing the micromechanics of interfaces between materials with different elastic constants and a better understanding of the plastic deformation and failure mechanisms of polymer glasses have fostered the development of new theoretical models of the adhesion between polymer glasses as well as stimulated improved experiments to measure this adhesion.

The important parameter obtained from an adhesion experiment is the critical energy release rate, $\mathcal{G}_c$, which is the energy necessary to grow a crack by a unit area. In glassy polymers, most of the energy necessary to propagate the crack is due to the plastic deformation near the crack tip. If this plastic deformation is to produce a reasonably large G_c, it must involve a certain volume in one, or rarely both, of the homopolymers bordering the interface. Although this volume of plastically deformed polymer is much smaller than the characteristic dimensions of the sample, therefore allowing linear elastic fracture mechanics to be used, it is much larger than the width of the zone of polymer mixing that defines the position of the interface, cf. Fig. 1. The magnitude of the plastic strain within this interfacial mixing zone rarely contributes significantly to G_c, rather the structure of the interface zone matters in permitting enough stress to be transferred across the interface to permit the growth of a large plastic zone, often a craze, on one side of the interface. The stress at which this zone can grow in the bulk polymer away from the interface will be of equal importance to the stress transfer across the interface itself; a low stress for such growth will ensure that the zone will involve a large volume of bulk polymer and thus produce a larger G_c for the same stress transfer across the interface.

Nevertheless, while the macroscopic measurable quantity is the fracture energy G_c, the actual failure of the interface eventually requires a molecular event such as chain pullout or chain scission within, or adjacent to, the plastic zone. This molecular event could be considered as being triggered by a critical force on the connector. The strategy for achieving optimum interface reinforcement may thus be formulated as follows: Provide the highest possible areal density of molecular connectors with a large critical failure force so that a large stress on the interface can develop in, or adjacent to, the plastic zone, thus permitting the plastic zone to grow as large as possible before the interface fails.

Typically the connector chains are di- or tri-block copolymers, random copolymers, or reactive chains which can graft at the interface. In practice these connector molecules must satisfy three requirements to be effective: (1) there must be a thermodynamic driving force for their segregation to, or reaction across, the interface between the two phases of the blend, (2) the kinetics of this segregation or reaction must be such as to ensure that a relatively high areal density of these connectors can form at the interface and (3) they must be mechanically effective in reinforcing that interface. While the first two of these conditions are controlled by thermodynamics and kinetics, respectively, the latter is related to the mechanical properties of the interfaces. Although the different connectors may seem quite different, their effect at the interface can be reduced to a general problem: the transmission of force through an entangled chain.

The examples shown in Fig. 2 are illustrations of various aspects of the problem: in the block copolymer case, shown in Fig. 2a, one of the two sides is typically weaker than the other so the problem can be reduced to the study of the weak side. If the connecting chains are grafted to a non-polymeric substrate or to a crosslinked network on one side, as shown in Fig. 2b, then the problem is more naturally defined. In the case of triblock copolymers, the two sides of the interface are inherently asymmetric through the different architecture of the chain but the reinforcement effect can be treated with the same formalism by de-

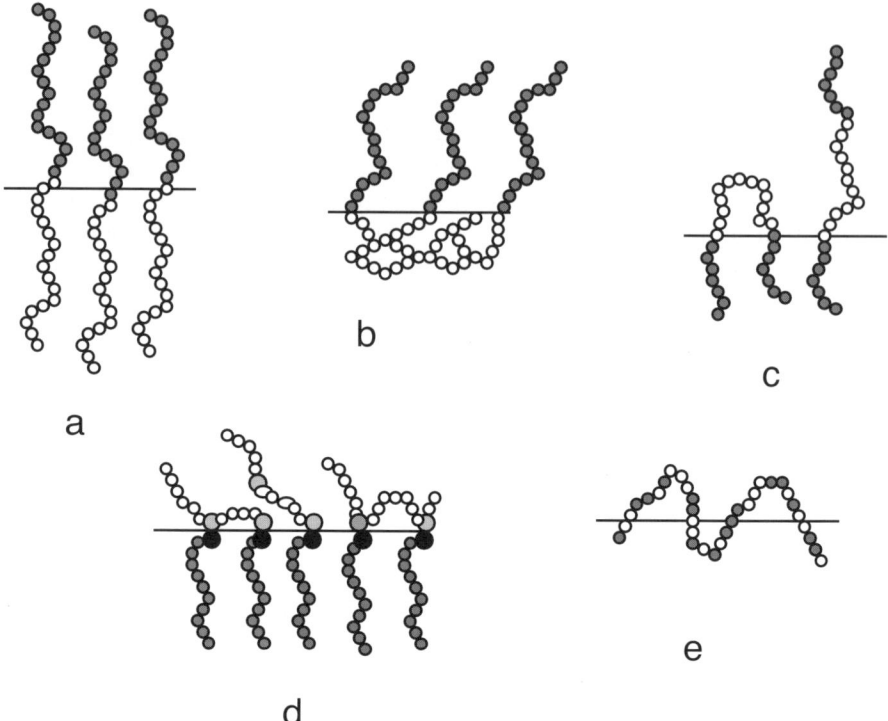

Fig. 2. Schematic of connecting chains at an interface. **a** diblock copolymers, **b** end-grafted chains, **c** triblock copolymers, **d** multiply grafted chain, and **e** random copolymer

fining a connector as a polymer strand crossing the plane of the interface; in that case each triblock copolymer chain contributes two connectors, as shown in Fig. 2c. In the case of random copolymers and grafted chains with multiple grafting points, the number of connectors can be much larger than the number of chains at the interface and the stress is transmitted by loops, as illustrated in Fig. 2d and 2e. Eventually the problem comes down to understanding what is the maximum number of connectors that one can have at the interface per unit area, and how effective these connectors are at transmitting the stress.

2
Experimental Methods

The recent progress in the understanding of the mechanisms controlling polymer-polymer adhesion, which are reviewed in this article, have become possible through the development of a certain number of experimental techniques and these can be divided in three categories:
– Synthesis of well-defined connector molecules that permit molecular design of the interface (Sect. 2.1).

- Molecular, surface and near surface analysis techniques combined with TEM observation (Sect. 2.2).
- A reliable mechanical test measuring the fracture toughness G_c of the interface (Sect. 2.3).

In order to understand the mechanisms that are responsible for the mechanical reinforcement of an interface by connector molecules and the main parameters controlling them, one must be able to work with well-characterized polymers and, ideally, a suitable technique for labeling these connector molecules must be developed to enable the fracture surfaces to be analyzed.

2.1
Synthetic Methods and Interface Formation

Diblock and triblock copolymers are usually synthesized anionically, which inherently limits the possible blocks and homopolymers. However, the anionic route gives very good control over molecular weight, molecular architecture and polydispersity and is ideally suited for a systematic study of the effect of connector molecules. Furthermore, the problems frequently encountered during the anionic synthesis have been ironed out for many common polymers and the synthesis, although tedious, can be considered a routine procedure.

Random copolymers can be produced through the much simpler free-radical route which, however, does not allow precise control over the molecular weight and yields a polydisperse polymer. It has been reported recently, however, that nearly monodisperse polymers can be obtained by controlled free-radical polymerizations [2–5].

One potential problem with conventional free-radical copolymerization is that the reactivity ratios of the two monomers tend to be different from one another [6]. On one hand this leads to non-random sequences of the monomers on a single chain (usually the product of the reactivity ratios is less than one so that there is a tendency to form alternating sequences) and, on the other, to substantial composition drift if the polymerization is carried out in bulk to high conversions. Random copolymers with a range of compositions as a result of composition drift may however be useful in practice, allowing a compositionally graded interface to be formed.

Reactive chains can be obtained by anionic polymerization, followed by attachment of a reactive end-group. This route yields nearly monodisperse polymers with functional groups at their ends, polymers that are very well suited for systematic studies. The synthesis can be quite elaborate, and, for longer chains, the completeness of the end-functionalization is difficult to verify. If the monodispersity and the control of the molecular weight of the polymer are not crucial or not possible, more common techniques can be used. They involve either the free-radical synthesis of a polymer incorporating a small fraction of reactive comonomers that will then be distributed along the chain, or the random functionalization of the polymer in the melt (using a free-radical initiator) after the polymerization. A common example of the former technique is the synthesis of styrene-maleic anhydride copolymer, while the second technique is typically used for the functionalization of polyolefins. Polymers produced by step-growth polymerization, e.g. polyamides, often have suitable end-groups for reaction at an interface if they are not end-capped.

For all these specialty polymers, deuterium can be used as a label on one or the other monomer. Deuterium labeling allows the use of techniques based on ion detection such as forward recoil spectrometry (FRES), nuclear reaction analysis (NRA) or secondary ion mass spectrometry (SIMS). If a high-resolution depth profile of the interfacial region is needed, neutron reflectivity can also be used. The main drawback of that approach is the cost of the deuterated polymers; while deuterated styrene and methyl methacrylate are expensive but commercially available, other monomers need to be synthesized and the cost can be quite prohibitive.

An important problem when testing polymer interfaces is obtaining a molecular structure at the interface that closely matches what would be expected in a realistic blend, or at least a well-controlled organization that can be predicted by the tools of thermodynamics. There are two main classes of methods: The first allows the chains to organize themselves locally at the interface to achieve a metastable equilibrium. Such processes can include chemical reactions near the interface but no long-range diffusion. This method was used for all the tests with block copolymers at the interface and rests on the assumption that, even if the whole specimen has not come to a true global thermodynamic equilibrium, the interfacial structure has at least organized itself to achieve a local equilibrium, the structure of which can be predicted using self-consistent mean-field theory.

For this type of test, the two homopolymers are typically compression molded into 1–2 mm thick sheets. A solution of the block copolymer is then spun-cast on one of the two polymers. The solvent and the homopolymer substrate to receive the block copolymer must be chosen so that the surface of the homopolymer substrate is not dissolved during the spin-casting operation. After a drying step, the two slabs are welded with the block copolymer in between to form an interface. The welding temperature is chosen to be above the respective glass transitions of the two homopolymers and the two slabs are held under a moderate pressure for a time sufficient to allow the local organization of the block copolymer at the interface to reach a metastable equilibrium.

A second, seemingly less artificial, method would be to add a certain amount of block copolymer to one of the homopolymers and let it diffuse to the interface. This method has not been used to prepare fracture samples with deuterium-labeled block copolymers for several practical reasons: Dissolving 5% of copolymer in a sheet of 50×100×2 mm dimensions requires 500 mg of deuterated polymer while a similar interface can be obtained by spin-casting with approximately 5 mg. The time involved to achieve true diffusional equilibrium over millimeter-scale thicknesses is prohibitively long at typical welding temperatures and the presence of a background concentration of deuterium-labeled polymer in one of the polymer phases can potentially interfere with the measurement of the areal chain density of the copolymer directly at the interface.

2.2
Surface Analysis and Characterization

Finally, the last essential experimental tool for the molecular understanding of adhesion is the ability to measure the areal density of connector chains actually present at the interface during the mechanical test. This measurement is done typically on the fracture surfaces after the fracture has occurred and requires reasonably planar interfaces for the use of quantitative surface analysis techniques. If at least parts of the connector molecules are deuterated, ion beam techniques, when available, are the most useful and the most quantitative [7]. Forward recoil spectrometry (FRES), also called elastic recoil detection, has the advantage of being very quantitative for the detection of deuterium [7, 8]. It has a depth resolution of 800 Å and a range of approximately 1 μm. Alternatively, nuclear reaction analysis (NRA) [9–11] and time-of-flight FRES [12] can be used and give an improved resolution at the expense of time necessary for achieving acceptable statistics, since both these higher resolution techniques have a lower sensitivity than FRES. A low-energy FRES technique [13] can also be used to achieve better depth resolution than the standard version with excellent sensitivity; however, this method suffers from a decreased range of profiling and increased susceptibility to artifacts due to surface roughness. Under normal circumstances one does not require better depth resolution than FRES since the total integral excess of the deuterated component at the interface can be analyzed to yield its areal chain density. Thus FRES is normally the technique of choice for deuterium-labeled polymers. Dynamic secondary ion mass spectrometry (SIMS) is more difficult to make quantitative than FRES but has a significantly better depth resolution (~100 Å) and can be used to depth profile elements other than deuterium (such as oxygen or nitrogen) [14]. With any of the ion beam techniques care must be taken to avoid artifacts due to radiation damage [15]. Radiation damage can result in loss of monomers [e.g. in poly(methyl methacrylate], depolymerization occurs and the monomer diffuses out of the sample). Cooling the sample to liquid nitrogen temperatures using a cold stage is effective in preventing loss of labeled polymer. Diffusion and evaporation of monomer-sized fragments at these temperatures is so slow that, even though the radiation damage occurs, the nuclei remain at the depth they started and the depth profile is unaffected. Limiting the ion dose on any area of the sample by frequently moving the ion beam spot is also effective. Using these two methods, ion beam analysis can be used to determine depth profiles of deuterium-labeled components near the fractured interface of any polymer pair.

Clearly, deuterated molecules are expensive and not always a viable option for economic reasons. X-ray photoelectron spectroscopy (XPS) can be an alternative when a heavier atom such as nitrogen or oxygen is present on the connector chains (but not on the polymer on at least one side of the interface). XPS has a sensitivity that decreases exponentially with depth and thus a rather limited maximum depth (~50–100 Å in most polymers). For this reason, quantitative analysis to determine the areal chain density with XPS requires a very careful preparation of the samples so that the detected element on the connector chains is never far from the interface, as well as special precautions to avoid contamination of the fracture surface with the element being detected. For example, using oxygen as the "tag" element will be very difficult if a polymer on either side of the interface is liable to surface oxidation in air. Nevertheless, XPS has been used very successfully to investigate the chemical grafting of nylon 6 chains to maleated polypropylene [16].

If measurement of the areal density of connector molecules after fracture of the interface is not possible, it is possible to evaluate the areal density if the interfacial agent is deposited at the interface as a spin-coated thin film and no substantial diffusion away from the interface into the bulk materials is expected. In this case, a measurement by ellipsometry of the thickness of the film (spun using the same conditions) on a silicon substrate can be used to determine directly the nominal areal density of connector chains at the interface. In cases where the connectors can diffuse away from the interface during the annealing stage, however, this method will not give reliable results.

2.3
Mechanical Testing of the Interface and Fracture Mechanics

A reliable mechanical test to measure the adhesion of the interface is required. The standard method to quantify adhesion is to drive a crack at the interface between the two bulk materials and measure the critical energy release rate, G_c, to propagate such a crack. The implicit assumption made in most measurements of G_c is that the external work is dissipated in the plastic deformation of a small volume close to the crack tip.

A quantitative approach to the reinforcement of interfaces with block copolymers is necessary. Such an approach requires a way to evaluate the strength of the interface separately from any change in the morphology or microstructure of the blend. While classic fracture mechanics tests such as compact tension and double torsion could be used, the asymmetric double cantilever beam (ADCB) test used in the first such study [17] has been adopted by all successive workers to date. In this experimental geometry, shown schematically in Fig. 3, a wedge (usually a razor blade) is inserted at the interface. In some versions of the test the

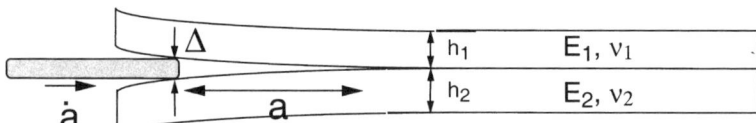

Fig. 3. Schematic of the asymmetric double cantilever beam geometry used for fracture toughness measurements

blade is pushed at a constant velocity while in others the crack is examined for a long time after the insertion. The length of the crack ahead of the tip of the wedge can be observed optically as long as one of the two homopolymer beams is reasonably transparent. The length of the crack is then directly related to the fracture toughness G_c provided that both the geometric parameters of the test and the elastic constants of the materials are known.

The mechanics of fracture along bimaterial interfaces have been studied extensively. Excellent reviews have been published [18]. The stress and deformation field near the tip of a crack lying along a bimaterial interface can be uniquely characterized by means of the complex stress intensity factor $K = K_1 + iK_2$. K_1 and K_2 have the dimension ($Pa\ m^{1/2-i\varepsilon}$) and are functions of the sample geometry, applied loading and material properties. $i = \sqrt{-1}$ is the imaginary number and ε is a dimensionless material constant defined below.

In experiments, it is easier to measure the energy release rate G which is related to K by:

$$G = C|K|^2/[16\cosh^2(\pi\varepsilon)] \tag{4}$$

where $\varepsilon = (1/2\pi)\ln[(\kappa_1/\mu_1 + 1/\mu_2)/(\kappa_2/\mu_2 + 1/\mu_1)]$, $C = (\kappa_1 + 1)/\mu_1 + (\kappa_2 + 1)/\mu_2$, and | | denotes the absolute value of a complex number $\kappa_i \equiv 3 - 4v_i$ for plane strain and $(3 - v_i)/(1 + v_i)$ for plane stress, where v_i and μ_i are Poisson's ratio and the shear modulus of material i, respectively. For example, for the PVP/PS system, where the Young's moduli and Poisson's ratios are: $E_2 = E_{PS} = 3$GPa, $v_2 = v_{PS} = 0.341$, $E_1 = E_{PVP} = 3.5$GPa, $v_1 = v_{PVP} = 0.325$, ε is found to be -2.8×10^{-3}. It should be noted that, for most polymer/polymer systems, ε is very small and can be set equal to 0. Note that, for the special case of a homogeneous material, $\varepsilon = 0$, K_1 and K_2 become the classical tensile and shear stress intensity factors K_I and K_{II}. In this case, the phase angle ψ is often referred to as the mode mixity at the crack tip, i.e.:

$$\psi = \tan^{-1}(K_{II}/K_I) \tag{5}$$

For the purposes of this review we will henceforth assume that $\varepsilon = 0$ for all interfaces between glassy polymers, so that simplification of Eq. (5) can be used.

In a homogeneous material, $\psi > 0$ implies that the crack has a tendency to propagate into the material below the crack line, whereas if $\psi < 0$, the crack has a tendency to be deflected into the material above the crack line [19]. For bimaterial cases, the direction of crack deflection depends on the details of the failure and deformation mechanisms of the interface and those of the bulk materials above and below the interface as well as on the phase angle. However, ψ can often be used as a guide to roughly predict the crack propagation direction.

Since G is a real number, it alone cannot fully characterize the crack tip field. In order to specify K, which has two real components, the phase angle ψ defined by Eq. (5) is also required. Using this definition, a point in the (G, ψ) plane is uniquely related to a point in the (K_I, K_{II}) plane. Following the concept of "failure locus" introduced by Rice [20] for a given phase angle ψ, the interface crack will start to propagate when the applied G reaches a critical value $G_c(\psi)$ which is assumed to be a material constant. This $G_c(\psi)$ is defined as the interface fracture toughness. Since G_c depends on ψ, the interface fracture toughness is a curve in the (G_c, ψ) plane.

The development of a test specimen for measuring interface fracture toughness involves finding analytical or numerical solutions for G and ψ. G and ψ for the ADCB specimen were computed using a boundary element method which was reported in [21]. These numerical solutions showed that the approximate energy release rate, G, given by Creton et al. [22] is a good approximation. This value of G was obtained from the beam on an elastic foundation model proposed by Kanninen [23]:

$$G = \frac{3\Delta^2 E_1 E_2 h_1^3 h_2^3}{8a^4 \Lambda^2}\left[C_1^2 E_2 h_2^3 + C_2^2 E_1 h_1^3\right] \qquad (6)$$

where $\Lambda \equiv C_1^3 E_2 h_2^3 + C_2^3 E_1 h_1^3$ and $C_i = 1 + 0.64(h_i/a)$

Figure 4 provides the G and ψ curves for the special case of $E_1 = 2E_2$. All G curves are approximately independent of the aspect ratio h_2/h_1. However, the phase angle ψ, which represents the shear/tensile mode mixity ahead of the crack tip, is positive for $h_1 = h_2$ and becomes negative as h_2 becomes larger than h_1. This gives an easy practical way to control ψ experimentally. Note also that the phase angle ψ is practically independent of a/h_1 if $h_2 \leq 2h_1$ and $a/h_1 \geq 3$. For a thickness ratio $h_2/h_1 > 2$, the phase angle is approximately a constant for $a/h_1 \geq 5$.

As pointed out in the preceding paragraphs, the first important problem to be addressed when using the DCB test on a new experimental system is to evaluate the phase angle ψ of the propagating crack. Early experiments by Brown showed that very different G_c results were obtained for the same interfacial molecular structure if the relative stiffness of the two beams composing the DCB sample were modified [24]. This effect is now understood as due to a strong dependence of G_c on the phase angle ψ of the propagating crack. Crazes were nucleated from defects in the interface ahead of the crack tip at an angle of 45° and grew into the polymer phase with the smallest crazing stress for positive values of the phase angle ψ. (Positive values of ψ correspond to a tendency for the crack to want to propagate into the less craze-resistant polymer phase.) These crazes grew wider as the crack approached and caused energy dissipation that resulted in a large increase in G_c over its value if the phase angle was negative. Over the range of phase angles from 0–10°, G_c was both a minimum and independent of the phase angle. In what follows we take values of G_c measured under these conditions of negative ψ as the true measure of interfacial adhesion.

The phase angle of the propagating crack can be altered by changing the stiffness of the two beams (which are typically the two homopolymers) either by changing the elastic parameters of one of the two beams (usually not practical) or by changing the ratio of the thicknesses of the two beams (the usual method). Although the phase angle ψ of the propagating crack can be calculated using the method of Xiao et al. [21] for a given geometry and elastic constants of the

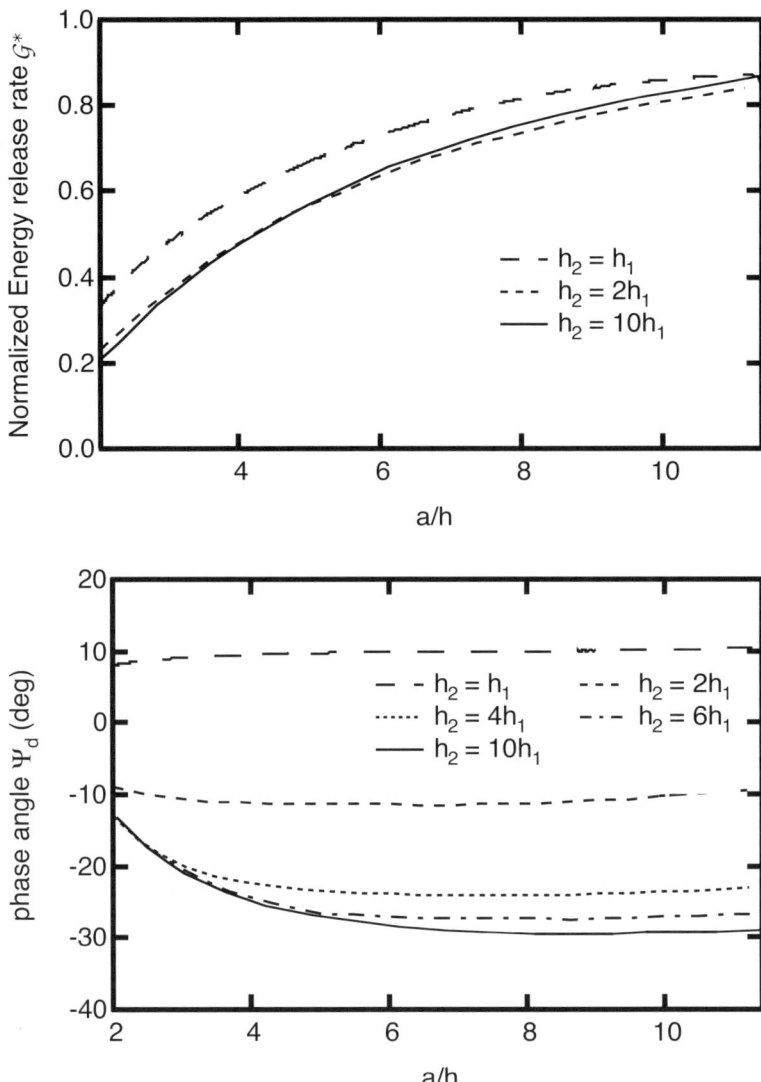

Fig. 4. Boundary element calculations of the phase angle and energy release rate for a crack propagating at an interface between polymer 1 and polymer 2. Example of the case where $E_1 = 2E_2$. **a** Energy release rate G^* (normalized by the value calculated with simple beam theory) and **b** phase angle, as a function of the ratio a/h_1 for different ratios of h_2/h_1. Data from [21]

beams, the effect of a change in phase angle on G_c cannot be simply predicted [25–27]. A way to avoid measuring the elastic properties and calculating the phase angle precisely is to measure G_c for a variety of relative thickness ratios for a series of interfaces with an identical interfacial structure, and to use then the ratio of thicknesses which gives the minimum value of G_c within a range of thickness ratios where ψ should be small but negative.

As a complement to the macroscopic mechanical test, Washiyama et al. [28] have developed a micromechanical test of the interface in a thin film that allows the deformation ahead of the crack tip to be observed directly using transmission electron microscopy (TEM). In this method, a DCB sample is microtomed edge on to produce a thin film (about 500 nm thick) that contains a single interface. That film is then deposited and bonded onto a copper grid using the method of Lauterwasser and Kramer [29] and subsequently strained, in such a way that the strain direction is perpendicular to the interface. The deformed film is then observed by TEM and G_c can be determined by measuring directly the thickness of the craze at the crack tip if the crazing stress is known. Although the results obtained with this method have to be treated with some caution, since it is not possible to control ψ always to be small and negative so as to match the conditions used for the DCB sample, they provide nevertheless valuable insights in the microscopic deformation and failure mechanisms that occur ahead of the crack near the interface.

To circumvent the limitations described above, Plummer et al. have used a different method, suitable for observation of plastic zones in bulk samples [30]. They embedded a DCB sample in a low viscosity epoxy resin with the razor blade in place. The crack tip was therefore maintained under stress while the resin was left to cure at room temperature. The sample was then trimmed for thin sectioning, stained by immersion in a RuO_4 solution, and microtomed in thin sections in the region of the plastic zone for observation by TEM. While this method gave particularly good results on ductile semicrystalline systems where a deformed thin film would not have been representative of the plastic deformation mechanisms taking place in bulk samples, it should in principle be applicable fairly generally.

3
Simple Connecting Chains Between Glassy Polymers

As pointed out in Sect. 1, the role of connector molecules in reinforcing interfaces between glassy polymers is to increase the maximum stress that can be sustained by the interface before failure. If the stress becomes higher than a certain critical level (which depends on the nature of the bulk polymers on either side of the interface), mechanisms leading to plastic deformation on a scale much larger than the molecular radius of gyration will be activated and the fracture toughness of the interface will increase dramatically. The main questions are:
1. What controls the maximum stress that can be sustained by the interface?
2. What controls the critical stress above which large-scale plastic deformation is observed?

These questions have been addressed mainly for the case where the connecting chains are diblock copolymers or end-grafted chains. In this case, each chain produces one molecular connection across the interface. The areal chain density of connectors is thus simply the areal chain density of molecules. We will illustrate the main features of the deformation and failure mechanisms with examples where the connectors are either diblock copolymers or end-grafted chains and then extend these results to the cases where each chain can contribute more than one molecular connection across the interface.

3.1
Main Experimental Results

Three homopolymer (diblock copolymer) phase boundary systems have been studied extensively: the system of polystyrene (PS) and poly(2-vinylpyridine) (PVP) reinforced with diblock copolymers of PS-PVP [22, 25, 28, 31–33], the system of poly(methyl methacrylate) (PMMA) and PS reinforced with diblock copolymers of PMMA-PS [17, 24, 34, 35] and the system of PMMA and poly(phenylene oxide) (PPO) reinforced by diblock copolymers of PMMA-PS [14, 36, 37]. Phase boundaries between PS and a crosslinked epoxy (XEp) were reinforced with carboxy-terminated PS chains whose –COOH ends reacted with either excess amines or epoxy to form a grafted brush at the interface [38, 39]. In a similar manner, interfaces between rubber-modified PS (HIPS) and XEp reinforced with grafted PS-COOH chains have been investigated [40].

Primary goals of these studies were to determine how the fracture toughness G_c of the interface between two immiscible glassy polymers depends on the areal density of connectors present at the interface and on the polymerization indices of each block of the diblock copolymer or of the end-functionalized chain. For that purpose, a series of diblock copolymers were synthesized by anionic polymerization to produce blocks with controlled polymerization indices where one block was usually labeled with deuterium. The grafted chains were produced in various lengths by anionic polymerization of deuterated styrene before reacting the macroanion with CO_2 to produce the –COOH end. One of the advantages of using these particular pairs of glassy polymers is that at least one of the homopolymers deforms plastically before the other by forming a stable craze ahead of the crack tip. Crazing is a relatively well-understood crack tip deformation mechanism and this understanding is very helpful in interpreting the fracture toughness of the reinforced interfaces.

Another advantage of most of these pairs is that the two polymers on either side of the interface have very similar thermal expansion coefficients in the glassy state. The similarity in thermal expansion coefficient means that only relatively small thermal residual stresses will exist after molding so that corrections to the strain energy release rate for these stresses [41] are not necessary.

The effectiveness of each diblock copolymer as a reinforcing agent for the interface was investigated in detail by varying the areal chain density Σ of copolymer deposited at the interface (as described in Sect. 2). A series of long symmetric PS-PMMA diblock copolymers (with N_{PS} varying between 420 and 2700) between PPO and PMMA were used to show that G_c depended just on Σ for long

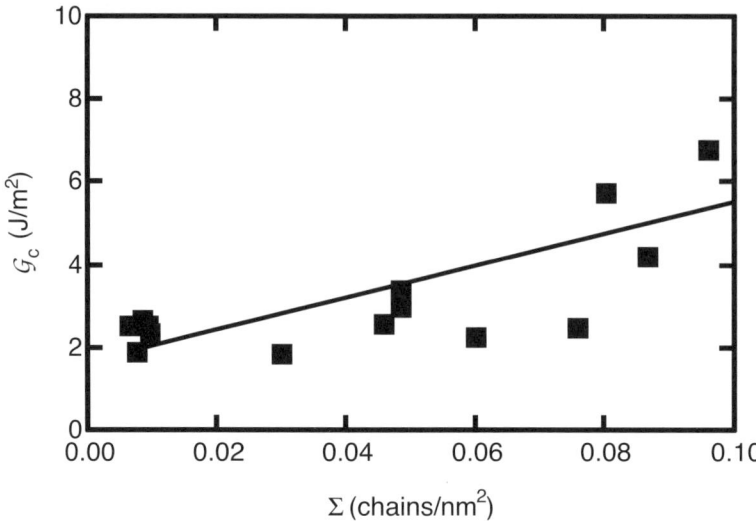

Fig. 5. G_c as a function of Σ for a 680-100 PS-PVP diblock copolymer at the interface between PS and PVP. Data from [31]

blocks [36]. To investigate the effect of reducing the length of one of the blocks, a series of dPS-PVP copolymers was synthesized where the degree of polymerization, N_{dPS}, of the dPS block was kept above 400 and that of the PVP, N_{PVP}, was varied from 49 to 870 [22, 31, 32]. To probe the effect of the dPS "block" length where the homopolymer was PS, end-functional dPS-COOH chains of various lengths N were grafted to the epoxy [38]. Three general mechanisms of interface failure were identified based on fracture toughness measurements and analysis of the fracture surfaces.

1. For values of N_{PVP} well below the average entanglement molecular weight of the PVP (N_{ePVP}), G_c typically increased linearly, but very slowly, with Σ, as illustrated in Fig. 5 for the 680-100 dPS-PVP copolymer [31]. (In all cases we will use a shorthand notation to describe the diblock copolymers. A 680-100 dPS-PVP copolymer is a diblock copolymer whose dPS block is 680 mers long and whose PVP block is 100 mers long.) Only a modest increase in G_c was observed from the bare interface value of 1.6 J/m^2 as Σ increased but nevertheless the slope of the G_c vs. Σ line increased with increasing N_{PVP}. The surface analysis results showed that 100% of the dPS block remained on the PS side, suggesting strongly that the PVP block was being pulled out from the PVP side of the interface. Chain pullout was also observed, this time in the PS, when a 150-140 dPS-PMMA copolymer was used between PS and PMMA. FRES results showed that all the deuterium remained on the PMMA side after fracture [34].

2. For slightly larger values of N_{PVP}, the measured fracture toughness showed a discontinuous increase at a given value of Σ, as illustrated in Fig. 6a for the

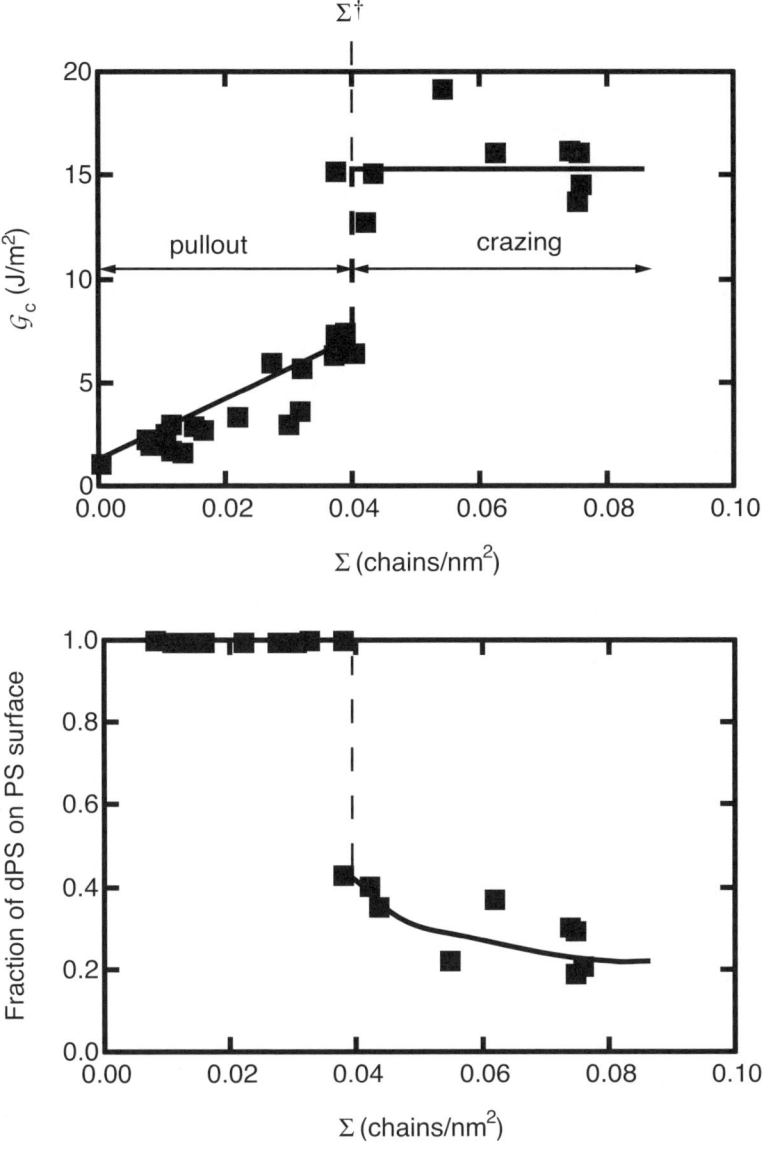

Fig. 6. **a** G_c as a function of Σ for a 580-220 PS-PVP diblock copolymer at the interface between PS and PVP. **b** Fraction of deuterium found on the PS interface after fracture as a function of Σ. Data from [32]

580-220 dPS-PVP copolymer, suggesting that a transition occurs between two different fracture mechanisms at that Σ [32]. The maximum value of G_c increased to approximately 15 J/m². The surface analysis results are presented in Fig. 6b and show conclusively that, whereas at low Σ all of the dPS remains

on the PS side, after the discontinuous jump in G_c, a much larger fraction of dPS is found on the PVP side, suggesting a different crack path and a change in failure mechanism.
3. For still larger values of N_{PVP}, or for long, grafted dPS chains, the G_c vs. Σ curve exhibits an apparent discontinuity in slope at $\Sigma \approx 0.03$ chains/nm^2. Below this value G_c increases very slowly with Σ but above this value the increase is much more rapid. The values of the fracture toughness are now much higher and can increase up to 100 J/m^2, as illustrated in Fig. 7a for the 800-870 dPS-PVP copolymer. The discontinuity in slope in the G_c vs. Σ curve was accompanied by a change in the locus of fracture (see Fig. 7b) so that, whereas the dPS block was found mostly, but not 100%, on the PS side at low Σ, implying chain scission near the copolymer joint, most of it remained on the PVP side at high Σ [22, 38]. This is due to the fact that, in this system, the weak point in the interface in the crazing regime was that between the homopolymer and the brush on the PS side.

One should note, however, that the discontinuity in the slope of the fracture toughness vs. Σ is not always accompanied by this change in the locus of fracture. When the connector chains on the crazing side of the interface are very long relative to the average molecular weight between entanglements of the homopolymer, the weakest point in the interface is no longer the homopolymer/brush but the joint area. Experiments using long dPS-PMMA and PS-dPMMA diblock copolymers between PS and PMMA have shown that copolymer scission typically occurs within just 33 repeat units from the copolymer joint in the low Σ region [14]. Subsequent experiments on the same system have shown that this result still holds even in the crazing regime [34, 36, 37].

The fracture toughness measurements, combined with the post-mortem surface analysis of the fracture surfaces, allow us to develop some simple micromechanical models to account for these three situations.

3.2
Fracture Mechanisms: Chain Pullout

For very short chains, the surface analysis results point clearly at a pullout of the shorter PVP block from the PVP homopolymer. For all block copolymers where $N_{PVP} < N_{ePVP}$, the FRES results showed that all the dPS was on the PS side after fracture. Although the same result could have been obtained, in principle, if the copolymer fractured exactly at the joint, in reality when fracture does occur near the joint, a small amount of dPS is always found on the "wrong" side. Also the results with PS-PMMA diblocks clearly demonstrate chain pullout for short chains.

This simple mechanism of chain pullout at the interface can be better visualized through a micromechanical pullout model [42] specifically developed for polymer glasses. Similar pullout models have also been developed for elastomer interfaces [43, 44].

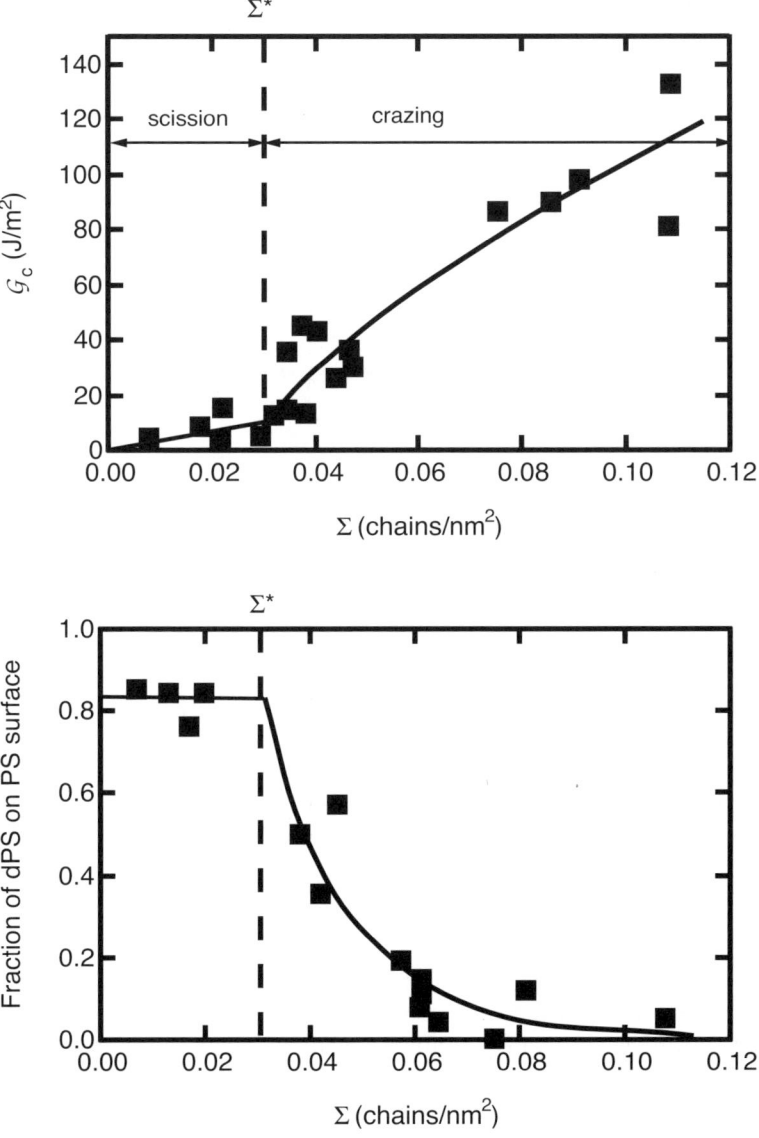

Fig. 7. a G_c as a function of Σ for a 800-870 PS-PVP diblock copolymer at the interface between PS and PVP. **b** Fraction of deuterium found on the PS interface after fracture as a function of Σ. Data from [22]

The chain pullout model developed by Xu et al. [42] is described schematically in Fig. 8. A chain, embedded in the polymer, is pulled at one end with a force f which is larger than the critical value $f^* = Nf_{mono}$ below which chain pullout cannot occur; f_{mono} is a monomer friction coefficient.

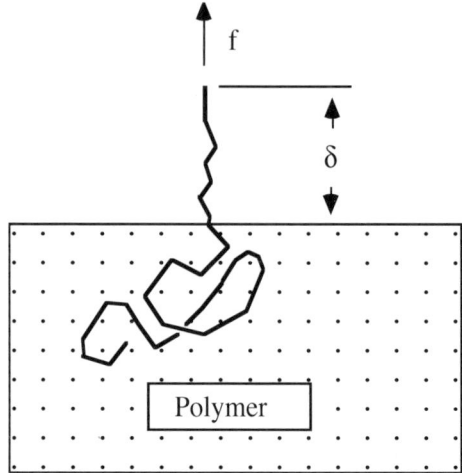

Fig. 8. A chain, embedded in the polymer, is pulled at one end with a force f which is larger than the critical value f^* below which chain pullout cannot occur; δ denotes the distance of the chain end from the homopolymer matrix

Let δ denote the length of the connector chain pulled out from the polymer by a force f as shown in Fig. 8. When $\delta = l$, where l is the total connector chain length, the chain is completely pulled out and the force f vanishes. Assuming that the chains are pulled out normal to the interface so that the tangential component of the pullout force is zero, then σ, the traction stress component normal to the planar interface, is related to the force f and Σ, the number of chains crossing a unit area of the interface, by $\sigma = f\Sigma$. σ is related to the rate of chain pullout $\dot{\delta}(t,x)$ and the remaining chain length $l - \delta$ by:

$$\sigma = b(l-\delta)(\dot{\delta}+\dot{\delta}^*) \qquad \dot{\delta} > 0 \tag{7}$$

where b and $\dot{\delta}^*$ are material constants. In this model, when the normal traction σ on the interface is less than $\sigma^* = b\dot{\delta}^* \equiv f_{mono}N\Sigma$, $\dot{\delta}=0$ and chain pullout cannot occur. For such a model f_{mono} can be considered to be a monomer static friction coefficient. Figure 9 shows a typical situation where this model can be applied to crack propagation.

The crack, semi-infinite in length, is assumed to propagate along the interface of two linearly elastic half spaces with a steady state velocity of $\dot{a}$ under small-scale yielding conditions, which implies that the region of pullout is small compared with typical specimen dimension. The interfaces are reinforced by chains which obey the pullout laws stated above. The steady state condition implies that all quantities are independent of time with respect to an observer moving with the crack tip.

Using this model, Xu et al. have demonstrated that for sufficiently slow crack growth rate, i.e. $\dot{a} \to 0$, the fracture toughness $G_c(\dot{a})$ is given by:

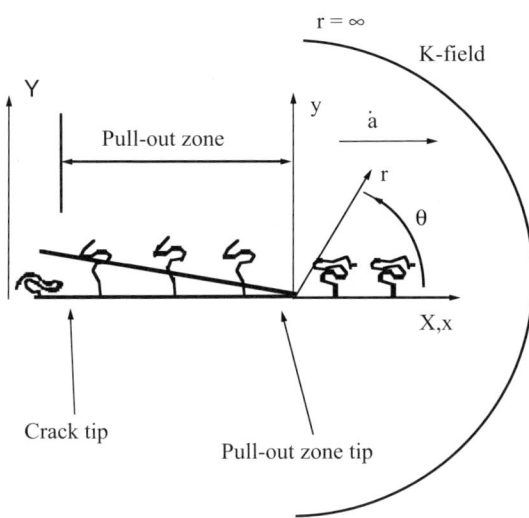

Fig. 9. A plane strain crack, semi-infinite in length, propagates along the interface of two linearly identical elastic half spaces with a steady state velocity of $\dot{a}$ under small-scale yielding conditions. The interface is reinforced by polymer chains. (x, y) is the cartesian coordinate frame attached to the moving crack tip. The pullout zone is defined as the region directly ahead of the crack tip in which interface opening is greater than zero but less than the critical value l, i.e. $l > \delta(x, t) > 0$.

$$G_c(\dot{a}) = G_c^o = \frac{\sigma^* l}{2} = \frac{f_{mono} \Sigma N^2 l_0}{2} \text{ for } \dot{a} \ll 0.36 \frac{\pi E^* \dot{\delta}^*}{\sigma^*} \qquad (8)$$

where $E^* \equiv E/(4(1-v^2))$ and $l = Nl_o$ where l_o is the length of a mer unit.

The main results of this micromechanical model in the quasi-static regime $(\dot{a} \to 0)$ can be compared with experimental results. The measured fracture toughnesses for the PS-PVP copolymers 625-49, 680-100 and 580-220 are shown in Fig. 10. In each case, the fracture toughness increases linearly from that of the bare PS/PVP interface while the slope of the line increases with the degree of polymerization of the block being pulled out. If the data for the three copolymers considered are replotted (see Fig. 11) as ΔG_c vs. $(N_{PVP})^2 \Sigma$ (where $\Delta G_c = G_c(\Sigma) - G_c(0)$, they fall on a single line consistent with a single value of the monomer friction coefficient. This value can be estimated by assuming that the only dissipative process is the viscous extraction of the PVP block where $\Delta G_c = f_{mono}(N_{PVP})^2 \Sigma(l_o/2)$ taking l_o, the PVP monomer length, to be 0.23 nm. The slope of Fig. 11 then yields a value of 2.5×10^{-11} N/monomer for f_{mono}. This value is, however, an overestimate as it does not take into account any viscoelastic dissipation near the interface. The most important point to emphasize, however, is that the maximum value of ΔG_c that can be obtained from this pullout mechanism is very low, less than 5 J/m². The reason is that the maximum displacement of the pullout zone at the crack tip is only of the order of 40 nm, the length of an unentangled connector chain.

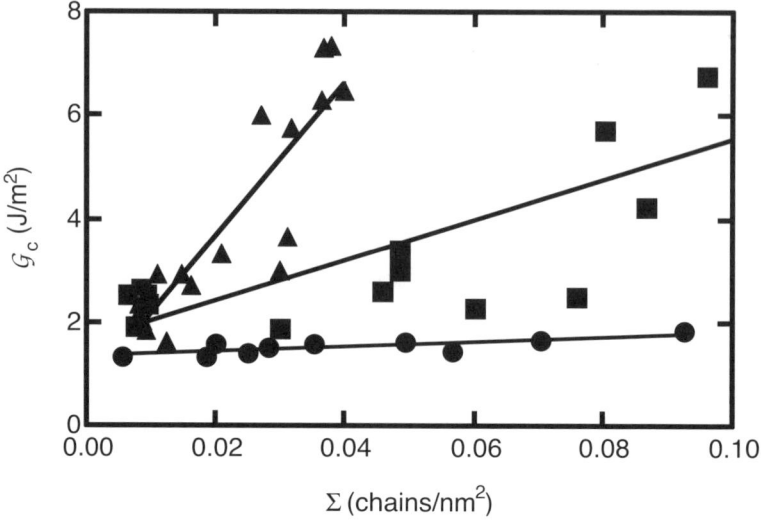

Fig. 10. G_c vs. Σ for PS-PVP diblock copolymers at an interface between PS and PVP. (▲) 580-220, (■) 680-100, (●) 625-49. Data from [31]

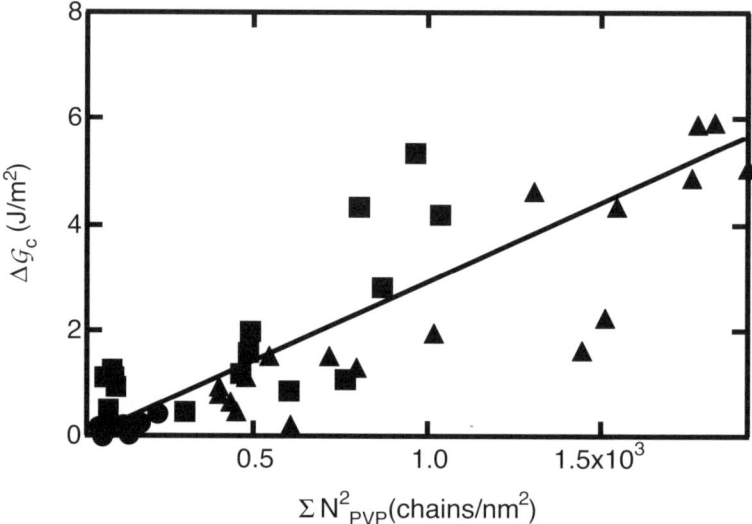

Fig. 11. ΔG_c as a function of $N^2_{PVP}\Sigma$ for diblock polymers 625-49 (●), 680-100 (■) and 580-220 (▲). Note that all points fall approximately on a single line

3.3
Fracture Mechanisms: the Transition from Simple Chain Pullout to Failure by Crazing

A better estimate of the static monomer friction coefficient can be obtained from the transition from failure by simple chain pullout to failure by crazing. For a given chain length, an increase in the areal density will result in a higher value of the stress that can be sustained at the interface without pullout occurring. The maximum stress that can be achieved will be determined either by the saturation of the interface (i.e. any additional block copolymer chain will form micelles and therefore not contribute to the load-bearing capacity of the interface), or by the plastic yield of one of the bulk homopolymers, the one with the lower yield stress. In the case of an interface between PS and PVP, both homopolymers deform plastically by crazing and have crazing stresses of 55 and 75 MPa, respectively [22]. Therefore one would expect the propagating crack to be preceded by a craze in the PS as soon as the load-bearing capacity of the interface exceeds 55 MPa. As shown in Fig. 6, such a change in fracture mechanism has indeed been observed [32] for the 580-220 dPS-PVP block copolymer at a value of $\Sigma = 0.04$ chains/nm^2 where G_c increases with a discontinuous jump from 5 J/m^2 to 15 J/m^2. Furthermore, the analysis of the fracture surfaces shows that all of the dPS remains on the PS side for $\Sigma < 0.04$ chains/nm^2 while most of it is found on the PVP side for higher values of Σ. From the value of Σ at the transition, which we designate $\Sigma\dagger$, one can extract another estimate of f_{mono}, provided that the crazing stress is known [32].

This gives for the PS-PVP case:

$$f_{mono_{PVP}} = \frac{\sigma_{craze_{PS}}}{N_{PVP}\Sigma\dagger} \qquad (9)$$

Numerically, this estimate gives $f_{mono} = 6.3 \times 10^{-12}$ N/monomer. This result should be compared with the estimate obtained with Eq. (8) which is four times higher. The discrepancy between the two results is due to the viscoelastic dissipation that was not taken into account in the energy calculation. Clearly, excess dissipation takes place near the interface even in the straight pullout regime.

Thus, in principle, determining f_{mono} from the pullout to crazing transition of connector chains can be used as a method to obtain a reliable friction coefficient in the glass. Unfortunately, in practice, the transition can only be observed in a narrow range of molecular weights. For the case of PVP, the transition is not observed for $N_{PVP} = 173$, where only pullout occurs before the block copolymer becomes saturated at the interface, while for $N_{PVP} = 270$, the chains break, i.e. chain scission rather than pullout occurs at low Σ [22]. For grafted dPS-COOH chains in pure PS, only pullout is observed for $N_{dPS} = 159$ [38] while a similar transition from chain pullout to crazing is observed for $N_{dPS} = 412$ in a matrix of high-impact PS (HIPS) [40].

For triblock copolymers with very short PVP end-blocks the situation is similar, as fracture occurs at low Σ by straight pullout of the PVP block. Experimen-

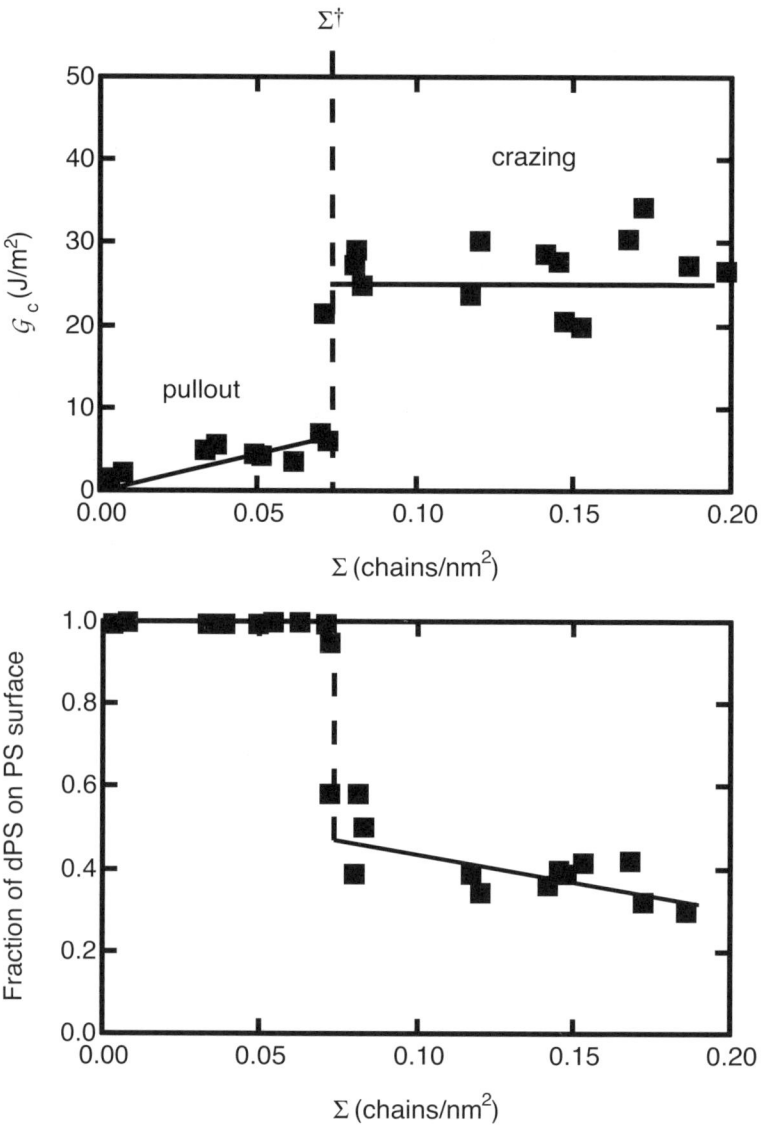

Fig. 12. Transition from chain pullout to crazing for a 90-580-90 PVP-dPS-PVP triblock copolymer at the interface between PS and PVP. At $\Sigma \sim \Sigma = 0.07$ chains/nm^2, G_c increases discontinuously from 8 to 25 J/m^2. **a** G_c as a function of Σ and **b** fraction of dPS on the PS side after fracture. Data from [46]

tally, one sees a sharp increase in G_c for $\Sigma > 0.07$ chains/nm^2 accompanied by a change in the locus of fracture, as illustrated by Fig. 12a and 12b. This is a signature for a transition from straight chain pullout to crazing. Using the static monomer friction coefficient calculated for the diblock copolymers of 6×10^{-12} N/mon-

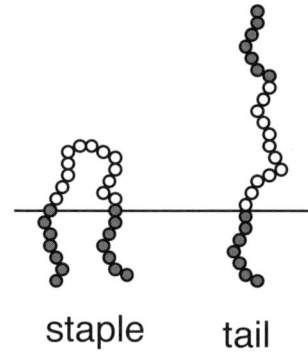

Fig. 13. Schematic of a triblock copolymer at the interface between PS and PVP showing the staple or the tail structure contributing respectively two and one connector per chain

omer, one can estimate a transition at $\Sigma = 0.05$ chains/nm² if each triblock contributes two connectors. However, the self-consistent mean-field simulation (SCMF) of Dai et al. [45] has shown that for such short PVP blocks the triblock can either be in a staple structure (contributing two connectors) or in a tail structure (contributing one connector), as shown in Fig. 13. Therefore, the high value for Σ^* of the 90-570-90 triblock can be accounted for by a high proportion (~60%) of the tail conformation at the interface.

3.4
Fracture Mechanisms: the Transition from Simple Chain Scission to Failure by Crazing

When N is further increased, the maximum force that is required to pull the chain out of the glass, i.e. $f_{mono}N$, will be higher than the force required to break a main-chain bond f_b. In this case one expects the maximum stress that can be sustained by the interface to be independent of N, and equal to $f_b\Sigma$. As long as this maximum stress remains lower than the crazing stress (or, more generally, the yield stress) of both the homopolymers, the chains will break without much plastic deformation, while when $f_b\Sigma > \sigma_{craze}$, a craze will form and propagate at the interface ahead of the crack, causing a sharp increase in the measured fracture toughness. For a given system, f_b and σ_{craze} are fixed so that the transition should occur at the same value of Σ, defined as Σ^*, provided that the connecting chain is long enough to avoid pullout. This transition has been observed in a variety of systems with connecting chains [22, 37, 38, 40] and, although the jump in G_c is not always clearly seen, there is always at least a sharp increase in the slope of the measured G_c vs. Σ plot.

As an illustration of this transition let us examine in more detail the case where the connector chains are dPS-COOH-grafted to an epoxy network. At low areal chain density and for $412 \le N_{dPS} \le 1478$, where N_{dPS} is here the degree of polymerization of the chains grafted to the epoxy, the measured fracture tough-

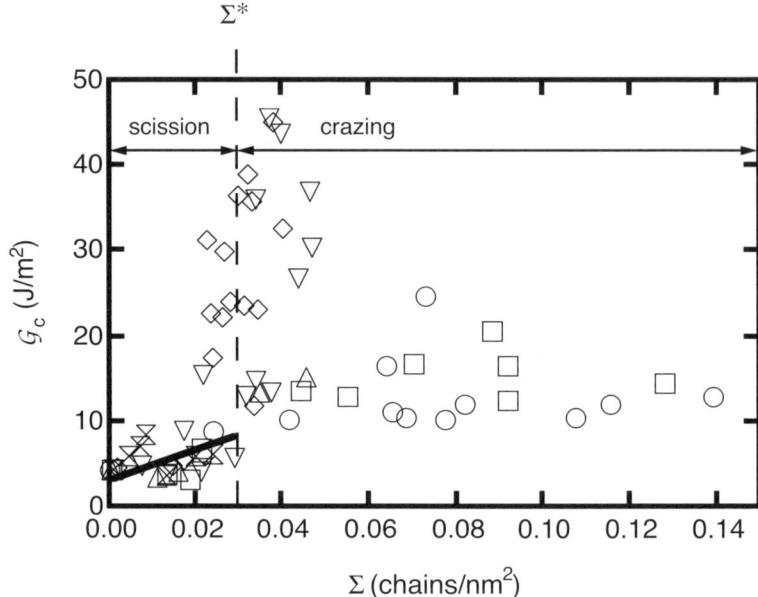

Fig. 14. G_c vs. Σ of interfaces between epoxy and PS reinforced with end-grafted dPS chains of various molecular weights. The transition from chain scission to crazing is clearly apparent for $\Sigma > 0.03$ chains/nm². $N = 412$ (○), $N = 535$ (▢), $N = 688$ (Δ), $N = 838$ (∇), $N = 1478$ (◇), $N = 1788$ (⌧). Data from [38]

ness is low and independent of N_{dPS}. All points fall on a single line (as shown in Fig. 14) and the surface analysis in that regime shows that the majority of dPS remains on the PS surface indicating a mechanism of scission of the dPS chain. When Σ is increased above a critical value of $\Sigma^* = 0.03$ chains/nm², G_c increases sharply indicating the onset of plastic deformation. As illustrated by Fig. 14, the transition from simple chain scission to crazing occurs at a unique value of the areal density of connecting chains and is indeed independent of the molecular weight of the connector. On the other hand, the value of G_c in the crazing regime is strongly dependent on N and this dependence will be discussed in the next section.

The value of Σ^* can be used to find a chain scission force $f_b = \sigma_{craze}/\Sigma^*$ if the crazing stress σ_{craze} is known, using reasoning similar to that used for the transition from chain pullout to crazing considered in the previous section. This fracture force should be characteristic of the fracture of a C–C bond and therefore not only independent of N, but also independent of the detailed molecular structure of the polymer. This appears to be indeed the case and the results obtained for several experimental systems [22, 37, 39, 40, 46] all yield a very similar value of $f_b \sim 2\times 10^{-9}$ N.

Using the experimental values of f_{mono} and f_b for the same polymer one can extract a prediction for the critical value N^* above which the chain is no longer pulled out but is fractured [32]. This critical degree of polymerization N^* is given by:

$$N^* = f_b/f_{mono} \tag{10}$$

which gives numerically $N^* \sim 350$ monomers for PVP. This value is larger than the N_{PVP} value (270) above which chain scission is first observed [22], suggesting that f_{mono} may increase when N is larger than N_{ePVP} (255) [32]. The increase in f_{mono} may be due to the onset of entanglement coupling between the connector chains and the homopolymer.

The same picture holds in principle for triblock copolymers at the interface. At low and medium areal densities of chains (i.e. when no multilayer of copolymer is formed at the interface), the results for G_c as a function of Σ which have been obtained for 90-570-90, 290-460-290 and 580-1620-580 triblocks show many similarities with the behavior of diblocks. The values of G_c and the surface analysis results for the 580-1620-580 triblock are shown in Fig. 15a and 15b. Similarly to the case of high molecular weight diblock copolymers, the strong increase in G_c for $\Sigma > 0.015$ chains/nm^2, accompanied by a change in the locus of fracture from the region of the PS/PVP interface to the region of the interface between the dPS block and the PS homopolymer, suggests a transition in fracture mechanism from chain scission to crazing.

This value for the transition should be compared with $\Sigma = 0.03$ chains/nm^2 obtained for the diblocks and is consistent with the hypothesis that the triblock organizes itself in a hairpin configuration at the interface and each chain contributes two effective crossings to the interface. If one defines Σ_{cross} as the areal density of strands crossing the interface, the results of the 580-1620-580 triblock and those for the 800-870 diblock can be compared (see Fig. 16). In this case the triblock behaves as two pseudo-diblock copolymers where the DP of the dPS block of the pseudo-diblock is equal to half that of the dPS block of the triblock and the DP of the PVP block of the pseudo-diblock is equal to that of the PVP block of the triblock.

However, while in principle one triblock is equal to two diblocks each with one half the molecular weight, for the purpose of transferring the stress to activate the plastic deformation mechanisms this analogy is no longer expected to hold for shorter middle blocks.

In the low Σ regime, where crazing is not active, the 290-470-290 triblock fails by chain scission. Clearly, in this case, the 460 middle block of the small triblock cannot be considered as two equivalent PS blocks of 230 since an interface reinforced by such chains would have failed by simple chain pullout. Intuitively, this result is not surprising as the geometry of pulling out a loop may require breaking a bond and thus should be more difficult than pulling out two linear chains.[1]

The transition from simple chain scission to crazing is also confirmed by TEM observations of interfacial cracks between PS and PVP in thin films [28] (as

---

1 The friction force of pulling a loop perpendicular to itself may also be higher than the friction force of pulling a chain along its contour.

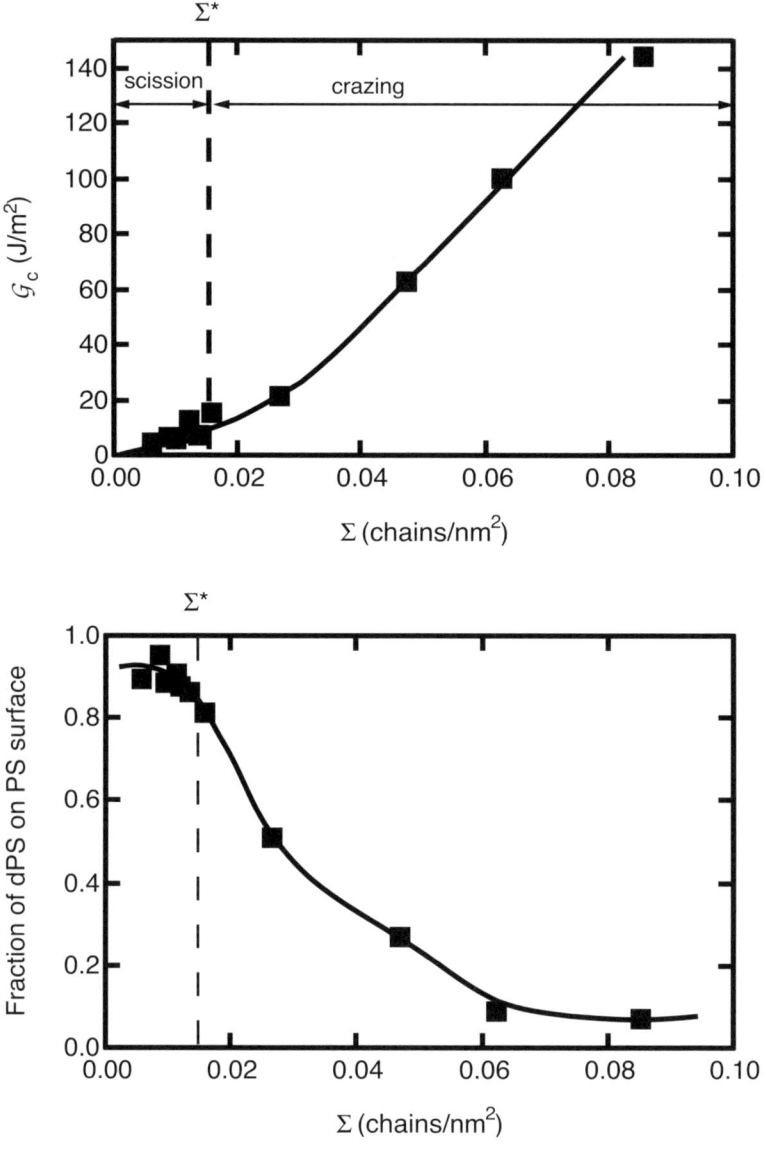

Fig. 15. a G_c as a function of Σ for interfaces between PS and PVP reinforced with 580-1620-580 PVP-dPS-PVP triblock copolymers. b Fraction of dPS on the PS surface after fracture. Data from [46]

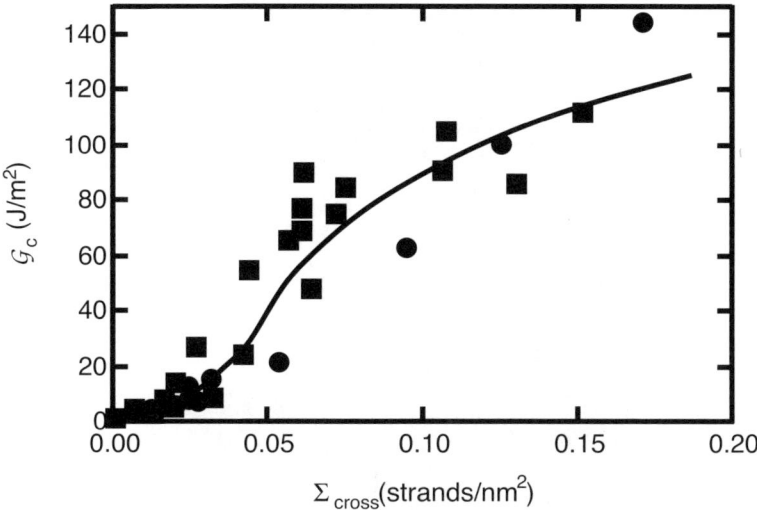

Fig. 16. G_c as a function of areal density of connectors Σ_{cross} for interfaces between PS and PVP reinforced with 800-870 dPS-PVP diblocks (■) and 580-1620-580 PVP-dPS-PVP triblock copolymers (●). Data from [22, 46]

described in Sect. 2.2). Two micrographs of an interfacial crack between PS and PVP are shown in Fig. 17. The interfaces were reinforced with an 800-870 diblock copolymer at two different areal densities. For $\Sigma > \Sigma^*$, a stable craze forms at the interface while this is no longer true for $\Sigma < \Sigma^*$. These TEM observations give us additional important information about the fracture micromechanisms: all the interfacial crazes for the PS-PVP system are completely formed on the PS side of the interface confirming that if a craze forms, it will do so in the bulk polymer with the lower crazing stress. This craze widening into only one of the homopolymers gives the craze tip the characteristic asymmetric shape shown in Fig. 18. The surface analysis of the DCB samples shows that the actual path of the crack must follow the interface because some deuterium, in all cases where crazing is active, is found on both fracture surfaces. Consistent with this result, the TEM observations show that the craze does not fail in its center but at the interface between PS and PVP, as shown in Fig. 19.

3.5
Fracture Mechanisms: Craze Growth and Stability

To achieve interfaces with the largest possible values of G_c it is necessary to select and then synthesize connecting chains that have the optimum values of N. One cannot simply synthesize connectors with the largest possible value of N and then construct an interface with a very large value of Σ. Such a strategy will not work because the largest value of Σ that can be achieved is a decreasing function

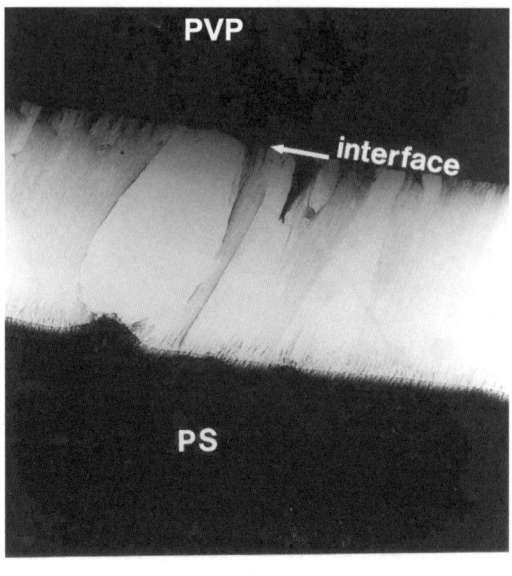

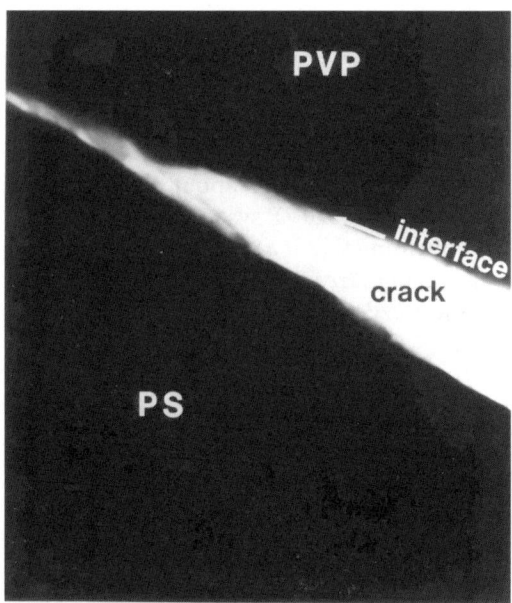

Fig. 17. TEM micrographs of an interfacial crack at the interface between PS and PVP [28]

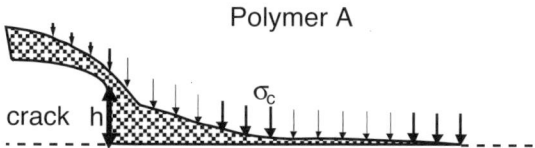

Fig. 18. Schematic of the craze formed ahead of a crack at an interface. Note that only the material with the lower crazing stress will be incorporated in the craze, which then grows in width in an asymmetric way

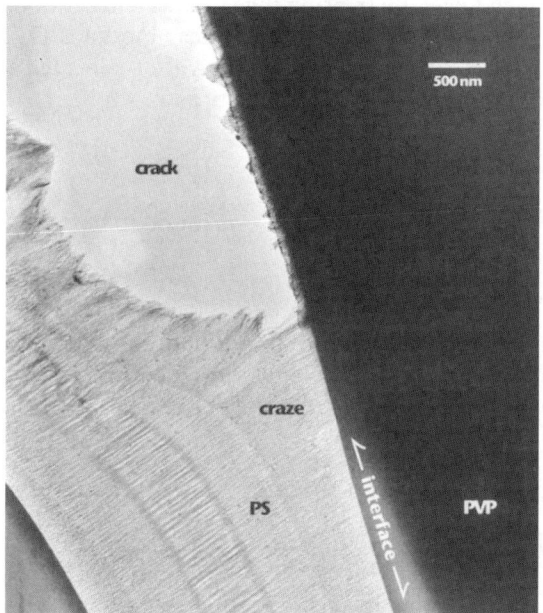

Fig. 19. TEM micrograph of a half craze at the interface between PS and PVP [28]

of N, both for block copolymers and for chains end-grafted at the interface by reaction. Some intermediate value of N will turn out to be optimal. To be able to predict this optimum, it is necessary to first develop a molecular understanding of the fracture mechanism of a craze along an interface reinforced with connectors of length N and areal density Σ and then to understand the linkage between N and Σ for the different kinds of connectors. The recent progress toward the first of these goals is presented below.

As seen from the previous discussions, crazing is the only deformation mechanism (for PS/PVP or PMMA/PS interfaces) that can lead to a significant increase in fracture toughness. The best current model for craze failure at a crack tip is that originally proposed by Brown [47]. Before Brown's work, the craze fibrils were always modeled as straight, parallel cylinders aligned along the normal to the craze surfaces and running from one craze interface to another. However, transmission electron micrographs and electron diffraction show the existence of short fibrils running between the main fibrils [48, 49]. A model for the microstructure of the cross-tie fibrils is shown in Fig. 20a. These "cross-tie" fibrils give the craze some small lateral load-bearing capacity so that they can transfer stress between the main fibrils behind the crack tip to the unbroken main fibrils ahead of the crack tip, as shown on Fig. 20b. Brown pointed out that this load-transfer mechanism allows the normal stresses on the fibrils directly ahead of the crack tip to reach the breaking stress of the entangled strands in these fibrils, even if the crazing stress σ_c (the stress needed for craze widening and growth) is much lower. By treating the craze as a highly anisotropic continuum with a longitudinal modulus C_{22} and shear modulus C_{66}, Brown showed that the tensile stress σ_{22} directly ahead of the crack tip located at the origin ($x = 0, y = 0$) has an inverse square root singularity as $x \to 0$, i.e.:

$$\sigma_{22} \cong K_{tip}(2\pi x)^{-1/2} \tag{11}$$

where K_{tip} is the stress intensity factor. K_{tip} is found to be

$$K_{tip} = A\sigma_c (C_{66}/C_{22})^{1/4} \sqrt{h} \tag{12}$$

where σ_c is the crazing stress, h is the half-width of the craze at the crack tip and A is a constant of order one. An excellent approximation for $\sigma_{22}(x)$, the full-field solution of Brown's model directly ahead of the crack tip ($y = 0, x$) was later obtained by Hui et al. [50, 51].

$$\sigma_{22} = \sigma_c(1 - \exp[-\pi x(C_{22}/C_{66})^{1/2}/h])^{-1/2} \quad x > 0 \tag{13}$$

It can be easily demonstrated that for $x \ll (C_{66}/C_{22})^{1/2}h$, Brown's asymptotic solution [Eq. (12)] is recovered with $A = \sqrt{2}$. Equations (15), (17) and (18) below are derived assuming that the craze is much wider than the fibril spacing, so that Eq. (12) is a good approximation for Eq. (13).

For the case of a craze at an interface, h is related to the continuum opening displacement δ at the crack tip by:

$$\delta = h(1 - v_f) \tag{14}$$

where v_f is the volume fraction of fibrils within the craze (≈ 0.25 for PS) [52, 53].

The ratio C_{66}/C_{22} depends only on the craze microstructure and the elastic properties of the fibrils, i.e. the spacing L between cross-tie fibrils which bridge between one main fibril and the next as well as the distance, d, between main fibrils. Estimates of C_{66}/C_{22} based on the micromechanics of the craze microstructure can be found in [54]. For crazes in PS, $L \approx 60$ nm and $d \approx 20$ nm, leading to an estimate for C_{66}/C_{22} of 0.02.

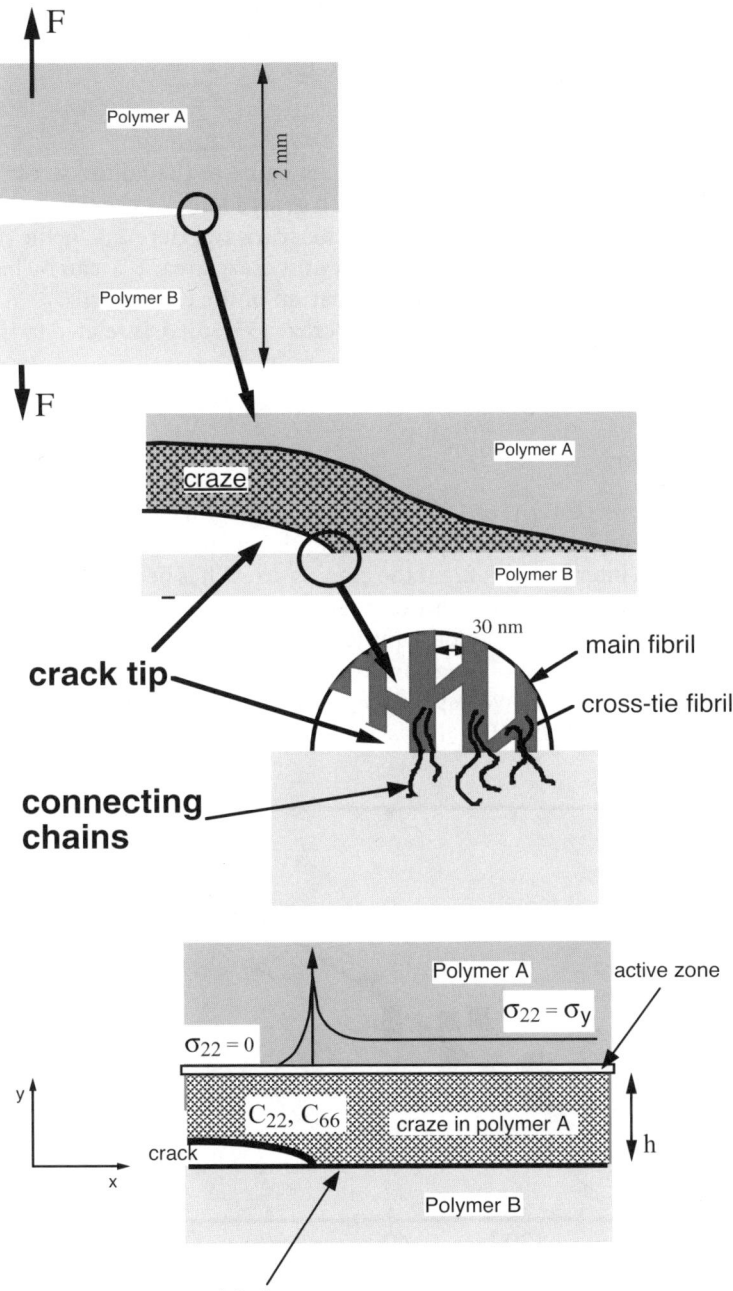

Fig. 20. a Schematic showing the various structures at different length scales near the crack tip within the craze. **b** Schematic of the stress field ahead of a crack tip (at $y = h$) when a craze is present

If one assumes that the craze is sufficiently thick so that $d \ll (C_{66}/C_{22})^{1/2} h$, the average tensile stress on the (continuum) fibril structure at a distance $D \approx d/2$ directly ahead of the crack tip is, according to Eq. (11),

$$\sigma_{fibril} = K_{tip}(2\pi D)^{-1/2} \tag{15}$$

To obtain an upper limit for G_c, σ_{fibril} is set equal to the failure stress $\sigma_{fibril} = \Sigma_{eff} f_b$ of the continuum fibril structure which gives a failure criterion for the critical crack opening displacement $\delta = \delta_c$ at the crack tip. Here Σ_{eff} is the number of effective connector strands per nominal unit craze area; Σ_{eff} can be less than Σ due to the scission of chains during formation of the craze fibrils.

If σ_c is assumed to be constant along the craze [55, 56], δ_c is related to the fracture toughness G_c by

$$G_c = \sigma_c \delta_c \tag{16a}$$

$$G_c = h(1 - v_f)\sigma_c \tag{16b}$$

The predictions of Eq. (16b) for the fracture toughness of interfaces where crazing occurs can be checked by measuring the craze width h at the crack tip along a thin film interface by TEM if the crazing stress has been previously measured [28]. A comparison between these results for G_c from TEM and those (G_c [MECH]) from asymmetric double cantilever beam samples with the same areal chain density of block copolymer is shown in Fig. 21. The agreement between the two methods is impressive, especially given the difference in constraint at the crack tip, and indicates that Eq. (16) is a suitable basis for further prediction. As

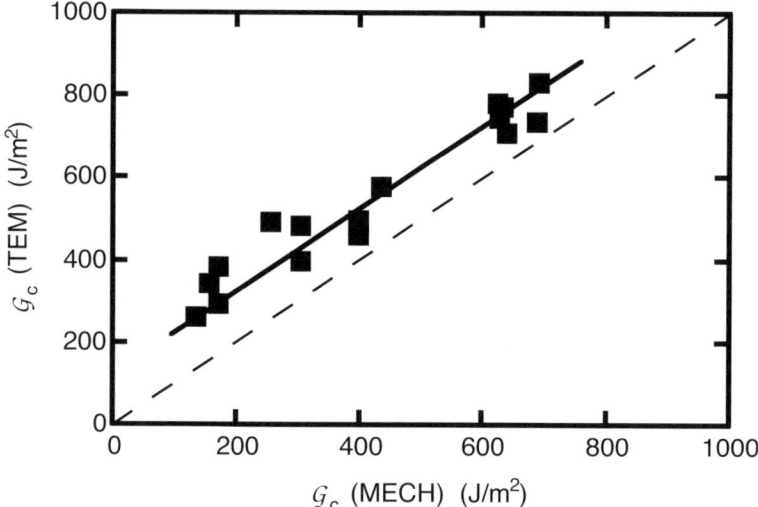

Fig. 21. Relationship between G_c [TEM] and G_c [MECH]. Note that G_c [TEM] is comparable to G_c [MECH] over the entire range of G_c examined and that G_c [TEM] ~ $2G_c$ [MECH] in this case. Data from [28]

further experimental evidence supporting Eq. (16b), a similar agreement between G_c [TEM] and G_c [MECH] has also been found in a different system where the value of h was measured on microtomed sections of thick samples [30]. In this case, h can be measured on the same sample used for the determination of G_c [MECH].

Using Eqs. (12) and (14–16) and the failure criterion $\sigma_{fibril} = \Sigma_{eff} f_b$, we find that

$$G_c = \frac{2\pi D(1-v_f)}{A^2}\sqrt{\frac{C_{22}}{C_{66}}}\frac{\Sigma_{eff}^2 f_b^2}{\sigma_c} = \frac{2\pi D(1-v_f)\sigma_c}{A^2}\sqrt{\frac{C_{22}}{C_{66}}}\frac{\Sigma_{eff}^2}{\Sigma^{*2}} \tag{17}$$

where the right hand equality in Eq. (17) follows from the definition $\Sigma^* = \sigma_c/f_b$. This result can be further simplified to yield

$$G_c = \frac{\pi d(1-v_f)\sigma_c}{2.88}\sqrt{\frac{C_{22}}{C_{66}}}\left(\frac{\Sigma_{eff}}{\Sigma^*}\right)^2 \tag{18}$$

where the 2.88 comes from choosing $D = d/2$ and selecting A such that the results for the stress in the fibril just ahead of the crack tip from the continuum solution matches that for simulations of a discrete model [51]. Note, however, that Eq. (17) cannot be even approximately correct for narrow crazes, i.e. weak crazes, as found either for Σ just greater than Σ^* or for crazes significantly weakened by disentanglement. The stress singularity of the continuum approximation [Eqs. (11) and (15)] is dominant only when $d << (C_{66}/C_{22})^{1/2} h$. For $h/d < 60$ (or $h/L < 20$), the full stress field given by Eq. (13) should be used for the continuum approximation, which means that G_c will not scale as Σ^2_{eff} for $G_c < 50$ J/m². In this case, G_c is given by:

$$G_c = \frac{\pi d(1-v_f)\sigma_c}{2\sqrt{\frac{C_{66}}{C_{22}}}\ln\left\{\left[1-\left(1.2\Sigma^*/\Sigma_{eff}\right)^2\right]^{-1}\right\}} \quad \text{for } \Sigma_{eff} > 1.2\Sigma^* \tag{19}$$

Equation (18) is recovered when

$$\left(\frac{\Sigma^*}{\Sigma_{eff}}\right)^2 << 1.$$

While theoretically the full-field continuum solution, Eq. (13), is an approximation for the stress in the last fibril, simulations that take into account the discrete nature of the craze and the detailed displacements of the craze/bulk interface due to fibril drawing near the crack tip [51, 54] show that it, and thus Eq. (19), are very good approximations. Note however that Eq. (19) is meaningless for $\Sigma_{eff} \leq 1.2\Sigma^*$: under these conditions the craze becomes so narrow that even the discrete model should be viewed with suspicion.

The treatment above assumes that within the craze there exists some effective areal chain density Σ_{eff} of connectors within the craze fibrils that must be broken with a force f_b. However, it is possible, and even likely, that the failure mechanism involves some disentanglement of connectors within the craze fibrils. Under these circumstances it may be more reasonable to find a failure stress σ_{fibril} for the craze fibril structure at the crack tip, a failure stress that is no longer given by $\Sigma_{eff} f_b$ but is a more complicated function of connector areal density Σ, connector degree of polymerization N, crack velocity $\dot{a}$, as well as temperature T. In what follows we assume that the craze microstructural parameters $\sqrt{C_{66}/C_{22}}$, v_f and d do not change with these other parameters even though in principle they will depend on rate and temperature. We define a standard crack tip displacement $\delta^\ddagger$ and a corresponding fracture toughness G_c^* as follows:

$$\delta^\ddagger = \frac{\pi d(1-v_f)}{2\sqrt{C_{66}/C_{22}}} \tag{20}$$

$$G_c^* = \delta^\ddagger \sigma_c \tag{21}$$

Equation (19) can now be rewritten as:

$$G_c = \frac{\delta^\ddagger \sigma_c}{\ln\left\{\left[1-\left(1.2\sigma_c/\sigma_{fibril}\right)^2\right]^{-1}\right\}} \tag{22}$$

In the limit of strong interfaces, Eq. (21) becomes simply:

$$G_c = \frac{\delta^\ddagger}{1.44} \frac{\sigma_{fibril}^2}{\sigma_c} = \frac{G_c^*}{1.44} \frac{\sigma_{fibril}^2}{\sigma_c^2} \tag{23}$$

It will prove instructive to use Eq. (21) to find the fibril failure stress under conditions, e.g. short connectors, where it cannot be assumed that all connectors break. For this purpose we can rewrite Eq. (21) to yield

$$\frac{\sigma_{fibril}}{\sigma_c} = \frac{1.2}{\sqrt{1-\exp(-G_c^*/G_c)}} \tag{24}$$

By fitting Eq. (23) to experimental data for strong interfaces where $\sigma_{fibril} = \Sigma f_b$ we can obtain a reasonable estimate of $\delta^\ddagger$ and of G_c^*, provided that the crazing stress of the more ductile material is known. These constants can then be used to predict the value of σ_{fibril}/σ_c for more complicated cases where either the actual areal density of effective connecting chains or the force to break a connector is not precisely known.

It should be noted that, while this model was developed with the structure of a craze in a glassy polymer in mind, a generalized form, such as Eq. (22), should apply equally well for polymers which do not craze but form a cavitational plastic zone ahead of the crack tip, such as semicrystalline polymers near a hard interface. Such a situation will be examined in Sect. 6.2.

Let us now consider the experimental evidence in favor of the crazing model presented above. The type of systems where the crazing model is expected to work best is that of strong interfaces with long connecting chains. In that case σ_{fibril} is known and is given by:

$$\sigma_{fibril} = \Sigma f_b \tag{25}$$

Figure 22 shows a plot of G_c vs. Σ/Σ^* for long PS-PMMA diblock copolymers at PPO/PMMA interfaces (circles) and dPS-PVP block copolymers at PS/PVP interfaces. The craze forms on the PMMA side of the PPO/PMMA interfaces and on the PS side of the PS/PVP interfaces. The dashed line represents Eq. (18) while the solid line represents Eq. (19). Both sets of data were taken from [22]. The data are superimposed on such a plot and are fit reasonably well by Eq. (18) with a value of $(C_{66}/C_{22})^{1/2} = 0.02$ that is also reasonable [57]. The full expression of G_c should be used for weak interfaces ($G_c <$ 100 J/m^2) otherwise the simplified Eq. (17) may be used.

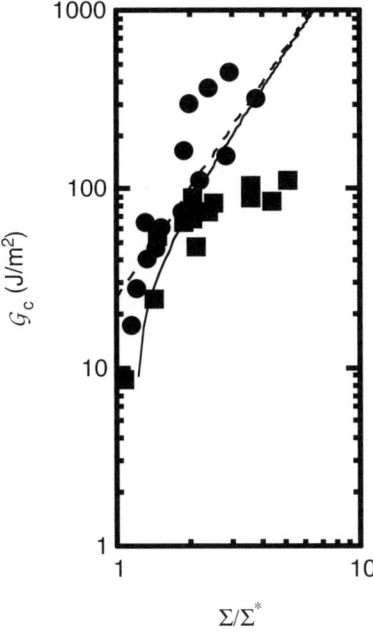

Fig. 22. G_c vs. Σ/Σ^* for interfaces between PPO and PMMA reinforced with PS-PMMA diblock copolymers (●) and for interfaces between PS and PVP reinforced with PS-PVP diblock copolymers (■). The *solid line* is a fit to Eq. (19) and the *dashed line* to Eq. (18). Data from [22, 36]

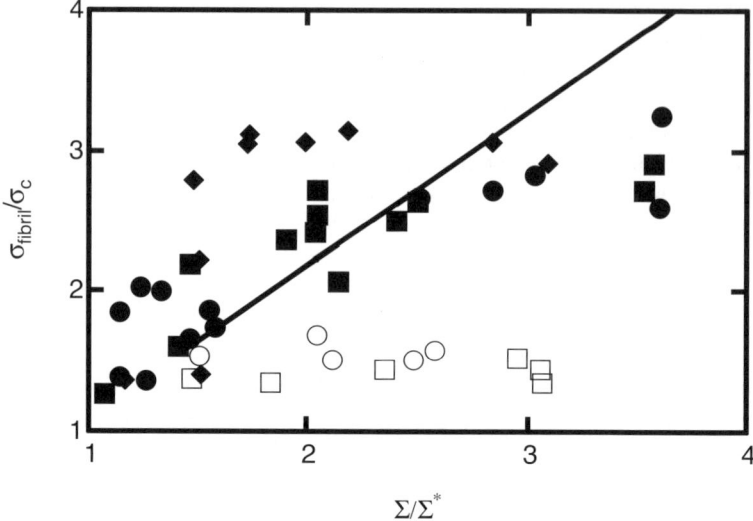

Fig. 23. σ_{fibril}/σ_c vs. Σ for various reinforced interfaces. (●) dPS-COOH, $N = 838$; (◆) dPS-COOH, $N = 688$; (■) PS-PVP, $N = 800–870$; (□) dPS-COOH, $N = 535$; (○) PS-PVP, $N = 510–540$. *Filled symbols* represent chains with $N \gg N_e$ while *unfilled symbols* represent shorter chains where disentanglement in the craze is expected. The *filled line* represents the value of σ_{fibril} one would expect from Eq. (25) while the *data points* are the value of σ_{fibril} obtained from the values of G_c and Eq. (24). Data from [22, 38]

By fitting the same experimental data to Eq. (23), we can extract a value for $\sigma^{\ddagger}$ and further apply the model to weaker systems where σ_{fibril} is no longer given by Eq. (25). In this case it is interesting to plot (see Fig. 23) σ_{fibril} [obtained from the value of G_c and Eq. (24)] as a function of Σ for different systems of connecting chains. Clearly, for the shorter chains ($N \sim 500$), σ_{fibril} still increases with Σ but no longer linearly. This indicates a certain amount of disentanglement of the polymer chain in the fibril so that the actual fracture stress is lower than the stress necessary to break every strand crossing the interface.

While in this case (short copolymer chains extracted from a matrix) one can interpret σ_{fibril} as the product of the areal density of copolymer chains by an average pullout force, this will no longer be the case if the value of Σ becomes ill-defined. One such case discussed below is the situation where the strength of the interface is no longer controlled by the molecular weight of the block but by that of the homopolymer.

3.6
Effect of the Homopolymer

The coupling between the connecting chain and the homopolymer occurs through the formation of entanglements. So far we have examined the effect of the molecular weight and areal density of the connecting chain on the fracture toughness, showing that the formation of entanglements with the homopolymer is crucial to obtain an interface with a high fracture toughness. However, one should keep in mind that, while the interface always fails very near the location of the connecting chains, tough interfaces generate a plastic zone several microns wide which develops therefore mostly in one or both of the homopolymers on either side of the interface. The plastic deformation properties of the homopolymers near the interface are therefore crucial to the formation of a stable craze and can have an important influence on the measured fracture toughness and on the fracture mechanisms. According to the model set out in the preceding section, the homopolymer present near the interface should have an influence mainly on two parameters: the crazing stress σ_c and the fibril stress σ_{fibril}.

In particular, the crazing stress does influence the onset of the pullout-to-crazing transition as well as the maximum width h_f of the plastic zone according to Eq. (19). From Sect. 3.4, the critical areal density for the onset of crazing, Σ^*, is given by:

$$\Sigma^* = \sigma_c / f_b \qquad (26)$$

where f_b is the force necessary to break a C–C main-chain bond. One expects therefore Σ^* to be proportional to σ_{craze} for vinyl polymers. This has indeed been observed for several systems. In certain cases it is also possible to inhibit the formation of a craze altogether by raising the value of σ_{craze} in such a way that Σ^* increases above the maximum attainable areal density of chains Σ_{max} [32].

For glassy polymers the crazing stress is usually weakly dependent on the sample preparation conditions and is a characteristic property of a given homopolymer. This is no longer the case, however, for semicrystalline polymers. In these systems it is difficult to separate the effect of cooling conditions on the coupling at the interface (possible crystallization of grafted chains and homopolymers in the same crystallite) from that on the characteristic size of the crystallites and on the degree of crystallinity. However, it is likely that a change in bulk mechanical properties of the homopolymer will influence the onset of the plastic deformation mechanisms.

A second category of polymers where a change in bulk properties is essential for good adhesion performance is that of rubber-reinforced plastics. In this case, as schematically shown in Fig. 24, one can have identical interfacial structures but very different bulk yield stresses due to the presence of rubber particles.

Two examples are shown in Figs. 25 and 26, where G_c and σ_{fibril} are plotted as a function of Σ. In the first case we compare interfaces between an epoxy and either PS or high-impact PS, where both interfaces have been reinforced with the same deuterated end-grafted chain, while, in the second case, we compare interfaces between polyamide 6 (PA-6) and either polypropylene (PP) or a PP-based alloy with a PP matrix and 70% EPDM rubber particles, where both interfaces have been reinforced with the same type of grafted PP chains. Two observations can be made:
1. The polymer with the lower crazing stress has a higher value of G_c.

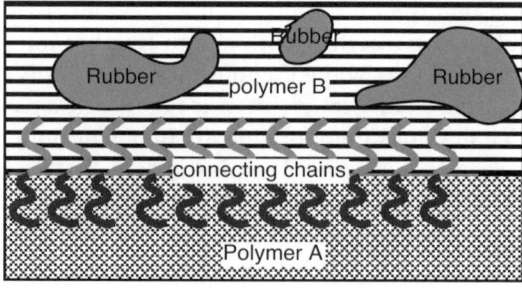

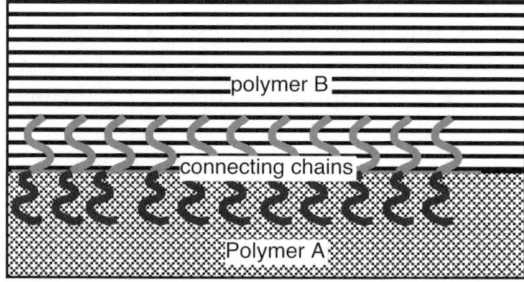

Fig. 24. Schematic of interfaces between polymers with an identical interfacial structure and different yield stresses: the example of rubber-reinforced polymers

2. For the PP system, the value of σ_{fibril} as a function of Σ is identical for both systems, implying that Eq. (24) applies to the plastic deformation mechanisms at the interface for both systems.
3. For the PS system, the value of σ_{fibril} obtained from Eq. (24) is unrealistically high (above the value of Σf_b which has to be an upper limit for σ_{fibril}). This implies that, in addition to a localized craze zone at the interface, other dissipative mechanisms, e.g. crazes nucleated from rubber particles well away from the interface, are active.

The other important influence of the bulk properties of the homopolymers is on the fibril stress, i.e. the craze stability. Even if, as pointed out above, the crazing stress for a glassy polymer does not vary much, it is well known that the molecular weight of the polymer has a profound effect on its fracture toughness [58].

In all the experiments described in Sect. 3, the homopolymer was a commercial grade high molecular weight polymer and was implicitly assumed not to be the limiting factor in obtaining a high fracture toughness at the interface. More recent experiments by Dai et al. [59] have shed some light on the role of the homopolymer in the entanglement coupling taking place at the interface. In these experiments, the polydisperse homopolymer PS normally used in most of the studies reported in this review was substituted with a series of monodisperse

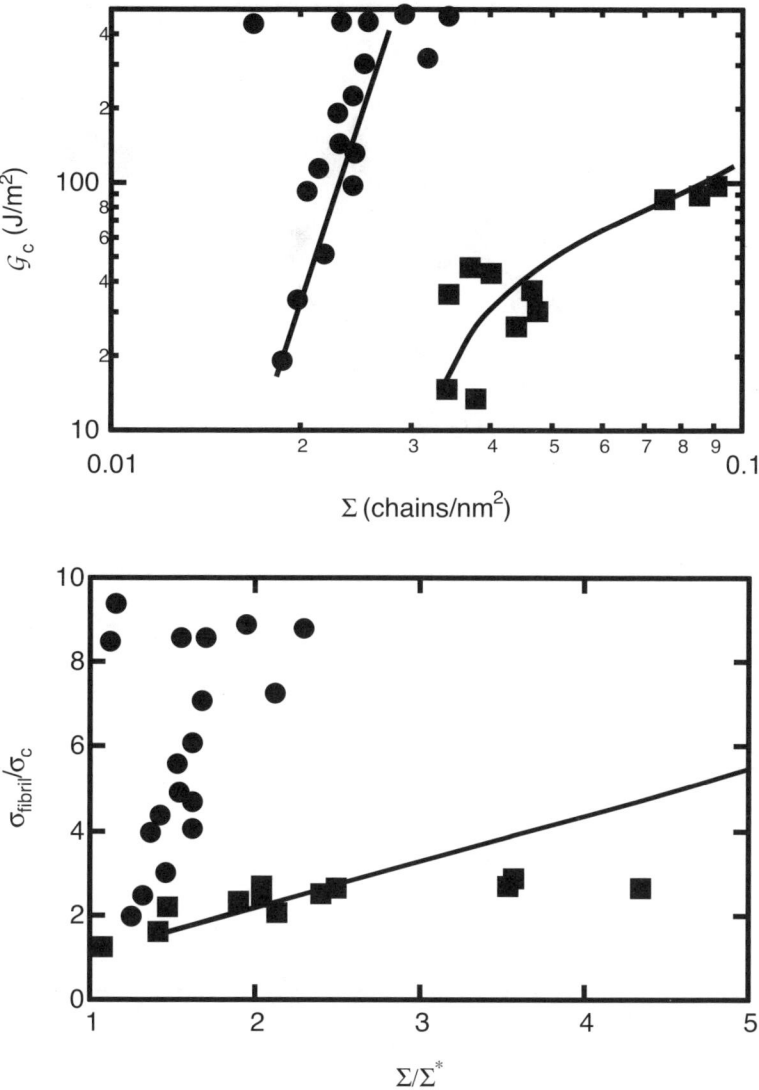

Fig. 25. a G_c vs. Σ and **b** σ_{fibril}/σ_c vs. Σ/Σ^* for an interface between PS (■) or HIPS (●) and epoxy reinforced by end-grafted PS chains ($N = 840$). The values of σ_{fibril}/σ_c are obtained from Eq. (24) and the *solid line* in **b** is obtained by setting $\sigma_{fibril} = \Sigma f_b$. Data from [38, 40]

PS (MPS) of different molecular weights as well as with a blend of high and low molecular weight polymers.

An interesting aspect in the understanding of the micromechanics of the interface is the fibril failure mechanism in the crazing regime. As pointed out in the previous section, two mechanisms are possible: chain scission or chain dis-

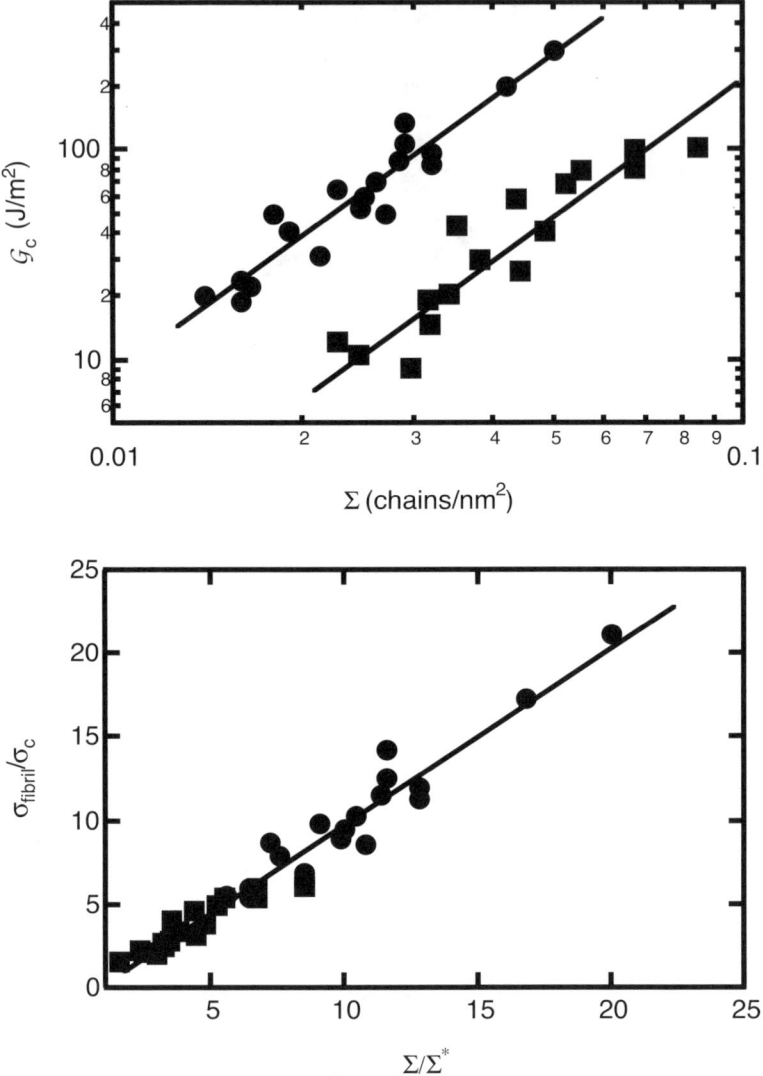

Fig. 26. a G_c vs. Σ and b σ_{fibril}/σ_c vs. Σ/Σ^* for an interface between PP (■) or a PP/EPDM blend (●) and PA-6 reinforced by end-grafted PP chains (M_n = 43,000). The values of σ_{fibril}/σ_c are obtained from Eq. (24) and the *solid line* in b is obtained by setting $\sigma_{fibril} = \Sigma f_b$. Data from [16, 102]

entanglement. The occurrence of one rather than the other is dependent on the ratio N/N_e where N is the degree of polymerization of the grafted chain. It is therefore instructive to consider first an interface reinforced with an areal density of copolymer, where the failure occurs well into the crazing regime for the experiments with the polydisperse homopolymer.

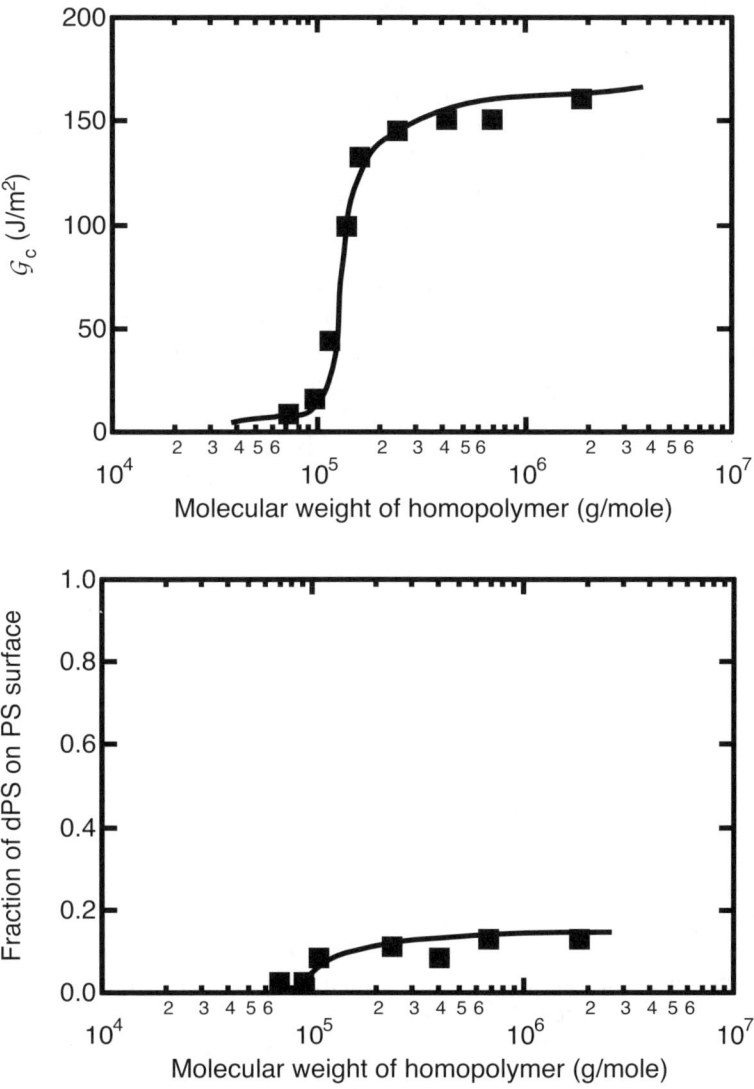

Fig. 27. a G_c of interfaces between MPS and PVP reinforced by 0.09 chains/nm² of 800-870 dPS-b-PVP block copolymer. G_c is plotted as a function of the molecular weight M of the homopolymer MPS. **b** Surface analysis for the same interfaces showing the fraction of dPS found on the PS side of the interface after fracture. Data from [66]

The fracture toughness of an interface between PS and PVP reinforced with 0.09 chains/nm² of 800-870 PS/PVP copolymer is shown in Fig. 27a as a function of the molecular weight of the MPS samples. G_c increases very sharply when M increases above 100 kg/mole and levels out at approximately 200 kg/mole. It is important to note that the value of G_c obtained for the polydisperse PS is consistent with its M_n rather than with its M_w.

The analysis of the fracture surface in Fig. 27b shows that most of the dPS is found on the PVP side of the interface even for high molecular weight MPS. This interesting result implies that the craze failure mechanism is controlled much more by the DP of the connecting chain than by the DP of the homopolymer. As one would expect, only when the molecular weight of the homopolymer drops below approximately 150 kg/mole (8–9 N_e for PS) does the molecular weight of the homopolymer affect G_c and the craze failure mechanism. These results have been qualitatively confirmed also on a shorter block copolymer (510-540).

The presence of a low molecular weight fraction in a high molecular weight polymer is known to embrittle it considerably by reducing the stability of its craze zone. Systematic experiments have been undertaken where a certain volume fraction of low molecular weight PS (M_n = 4000) has been incorporated into a high molecular weight PS (M_n = 670,000). This blend has then been used as the MPS for the fracture experiments and the results of G_c as a function of the volume fraction ϕ of the low molecular weight PS are shown in Fig. 28a while the fraction of deuterium found on the PS side is shown in Fig. 28b. The interface was in all cases reinforced with 0.09 chains/nm² of 800-870 PS-PVP to allow comparisons with the results of Fig. 27. Clearly, G_c decreases dramatically with increasing ϕ and the failure mechanism becomes increasingly chain disentanglement.

The results of Dai et al. can be analyzed with Eq. (24) and the measured fibril stress σ_{fibril}^{meas} can be plotted (see Fig. 29) as a function of the M_n of the homopolymer. This can be compared:
- to the predicted value σ_{fibril}^{calc} represented by a solid line given by $q\Sigma_{FRES}f_b$ where Σ_{FRES} is the areal chain density of copolymer measured using FRES.
- to the predicted value for the bulk PS areal density of strands represented by a dashed line (which varies with MPS because of chain ends).

These results can be interpreted as follows:
1. The maximum value of σ_{fibril} that can be achieved with a diblock copolymer remains lower than the value for the bulk homopolymer. This explains, therefore, why the measured G_c for interfaces reinforced with diblock copolymers remains lower than the bulk G_c of the homopolymers.
2. The difference between σ_{fibril}^{meas} and σ_{fibril}^{calc} gives a measure of the degree of pullout of the diblock from the homopolymer. When $\sigma_{fibril}^{meas} \approx \sigma_{fibril}^{calc}$, as for the 800-870 copolymer when the M_n of the MPS > 150,000, the fracture occurs mostly by chain scission according to the model of Sect. 4.1. On the other hand, for M_n < 150,000 for the 800-870 and at all molecular weights for the 510-540 copolymer, σ_{fibril}^{meas} is significantly lower than σ_{fibril}^{calc}, indicating that significant chain disentanglement from the homopolymer occurs.

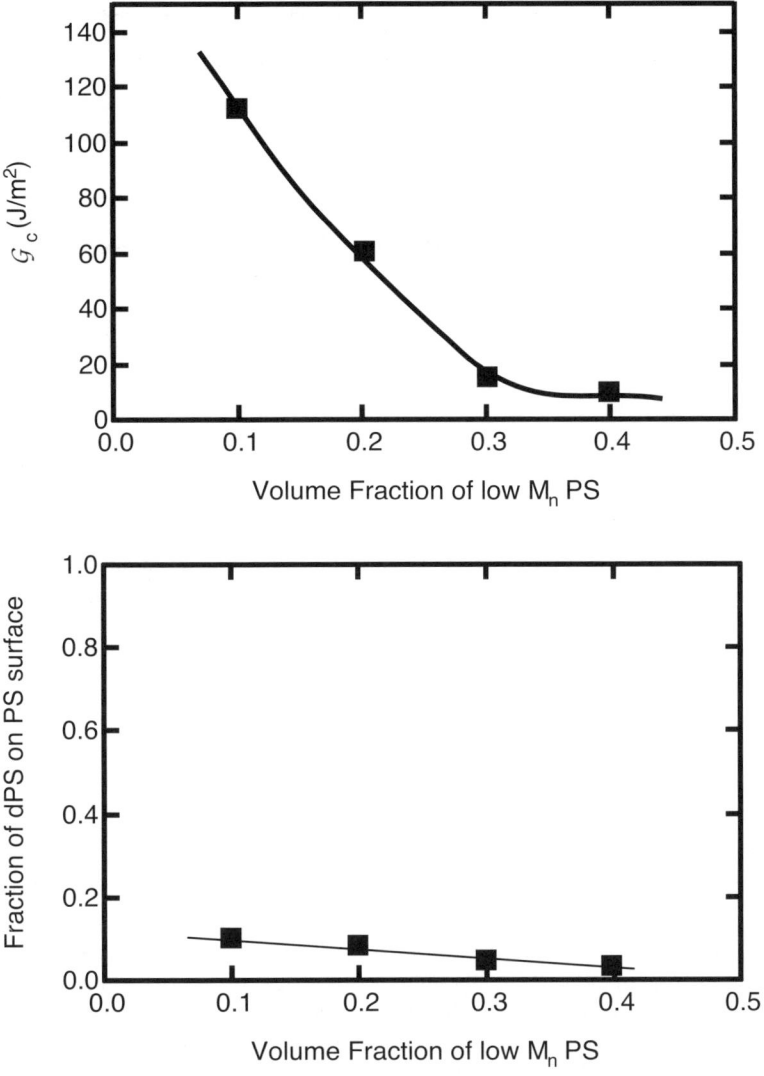

Fig. 28. a G_c of interfaces between a blend of high (M_n = 670,000) and low (M_n = 4000) molecular weight MPS and PVP reinforced by 0.09 chains/nm² of 800-870 dPS-b-PVP block copolymer. G_c is plotted as a function of the volume fraction ϕ of the low molecular weight MPS. **b** Surface analysis for the same interfaces showing the fraction of dPS found on the PS side of the interface after fracture. Data from [66]

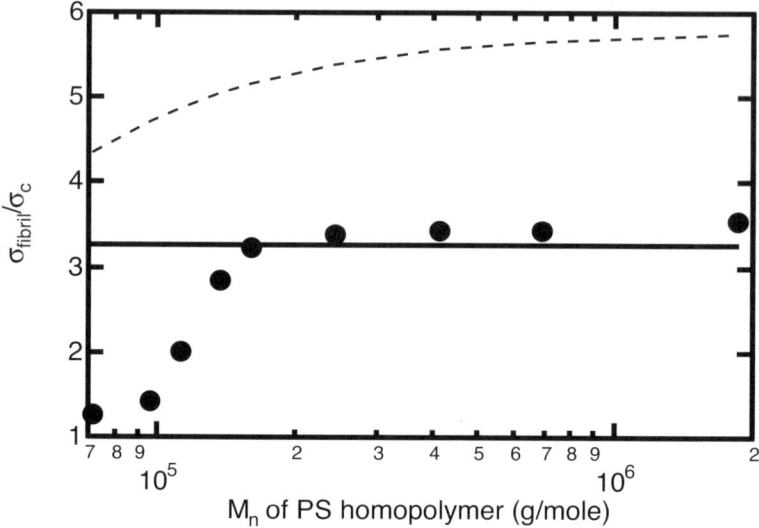

Fig. 29. σ_{fibril}/σ_c as a function of the molecular weight of the homopolymer MPS. The data points are obtained from the measured G_c values using Eq. (24). The *solid line* represents the predicted value obtainable with the measured areal density of copolymer chains at the interface Σ while the *dashed line* represents the predicted value for the bulk PS areal density of strands (which varies with MPS because of chain ends)

3.7
Velocity Dependence

Most of the experimental results presented above were obtained for the case of a steady-state crack propagating around 1–5 µm/s. The crack velocity in the DCB experiment performed with a wedge is controlled by the velocity at which the wedge is pushed to separate the sample. It is therefore possible in principle to do tests over a range of velocities. However, a few studies have been reported where the velocity of crack propagation has been investigated in a systematic way. The trend in these studies, illustrated by Fig. 30 in the case of a PS/PVP interface reinforced with a dPS-PVP 800-870 diblock copolymer, is however always one of increasing G_c with crack velocity [60].

An increasing value of toughness with crack velocity is commonly found in elastomeric systems; however, in this case, the increase in G_c is attributed to the viscoelastic losses in the bulk of the elastomer which increase with increasing deformation rate and decreasing temperature [61, 62].

For glassy polymers the same argument does not hold since, except in specific cases where secondary relaxations in the glass are active, the viscoelastic losses in the polymer, if any, should decrease with increasing crack velocity. However, we can still interpret the experimental results with Eq. (24) and note that the only two parameters that will vary with deformation rate are σ_c and σ_{fibril}. Clear-

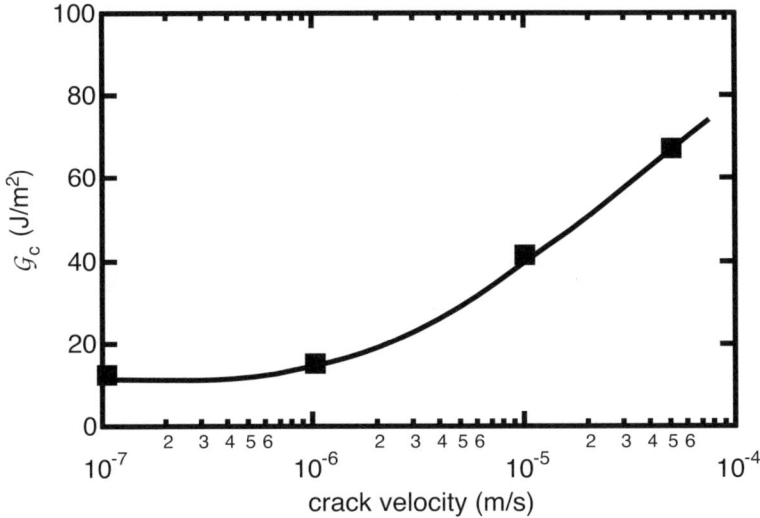

Fig. 30. G_c vs. crack velocity for the PS/PVP interface reinforced with 800-870 dPS-b-PVP diblock copolymer with $\Sigma = 0.04$ chains/nm^2. Data from [60]

ly, the crazing stress cannot decrease with increasing crack velocity but presumably the rate dependence of σ_{fibril} is stronger and dominates the behavior of G_c. One can envision three possible causes for this dependence:
1. In the disentanglement regime (relatively short chains in the fibrils) σ_{fibril} should increase with crack velocity because the disentanglement becomes more difficult.
2. However, in the scission regime, this argument does not hold and one would have to introduce a kinetic theory fracture argument (the fracture stress of a bond depends on the rate of deformation).
3. In the specific case of polystyrene, the propagation of an interfacial craze is often accompanied by oblique crazes in the bulk PS ahead of the main craze. The number and length of these crazes (which dissipate energy) can depend on the crack velocity [63].

3.8
Effect of an Elastomeric Midblock Within the Connecting Chains

The results described above were obtained in systems where both the homopolymers and the connecting chains were in the glassy state. It is well known that small amounts of elastomer can toughen glassy polymers by initiating crazing, so Brown et al. [64] examined the effects of using connecting chains with an elastomeric center block. They used PS-PB-PMMA and PS-PEB-PMMA triblock copolymers, (where PB is polybutadiene and PEB is polyethylene-co-butylene) as connecting chains between PPO on one side and PMMA or styrene acryloni-

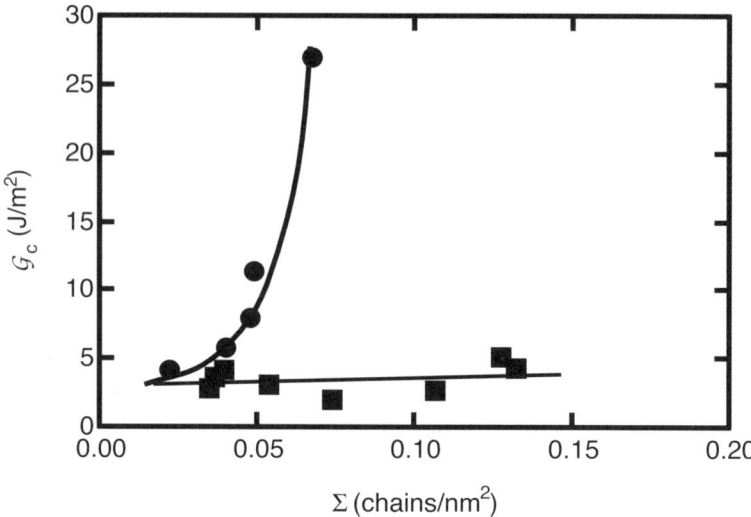

Fig. 31. G_c vs. Σ for interfaces between PMMA and PPO reinforced by: (●) 1400-1400 PS-PMMA symmetric diblock copolymers and (■) 560-1650-625 PS-b-PB-PMMA triblock copolymers. Data from [64]

trile (SAN) on the other. The elastomer was expected to be immiscible with all of the glassy polymers and therefore to form a separate microphase. A range of copolymers was used, with N_{PS} between 560 and 1000 and N_{PMMA} within a similar range, so failure by chain scission rather than pullout was expected.

The failure results showed the remarkable effect, illustrated in Fig. 31, that the connecting chains were very ineffective at toughening the interface, and in fact often decreased the toughness from that of the interface with no connecting chains. To test if this surprising result was caused by the elastomeric state of the center blocks in the connecting chains some fracture tests were made at –60 °C, which is below the glass transition temperature of the elastomers. For comparison, low-temperature measurements were also made on samples coupled by PS-PMMA diblocks of similar DP. Cooling the triblock-coupled samples increased their toughness by large factors while the toughness of the diblock-coupled sample did not change significantly. At the low temperature the triblocks were effective coupling agents that presumably failed by chain scission. Clearly, it is the existence of the extremely fine elastomer phase at the interface that suppresses the mechanical coupling at room temperature.

The experimental results described above can be explained within the basic fracture mechanism map after detailed consideration of the processes necessary to generate a craze at the interface. The criterion $\Sigma f_b > \sigma_{craze}$ is a necessary condition for the formation of stable craze fibrils. However, it is not sufficient for the formation of a craze at an interface. Craze initiation is believed to occur by a meniscus instability process that happens within a yield zone (an active zone) at a crack tip. This yield zone blunts the crack tip and averages out the stress singularity that would occur in a purely elastic system. It would seem likely that the length of this yield zone needs to be in the order of at least two fibril diameters for the meniscus instability process to occur. At an interface between two materials this yield zone must occur partly in the mixed layer of material. Hence the mixed layer has to sustain a stress a little above the polymer yield stress (because of the constraint), perhaps 100 MPa in a zone of length 30 nm, to start the crazing. The opening of this zone would be about 2 nm so it could occur within the mixed layer. The formation of this zone can only lead to stable crazes if the fibrils, as they form, are stabilized by sufficient entangled coupling chains, hence the crazing criterion given above. If the mixed layer contains elastomeric material then it is liable to pull apart at stresses well below the glassy yield stress, and so the crazing process cannot start. Instead, during crack propagation, the highly loaded elastomer-containing chains from the connector triblocks would be broken successively with little energy dissipation.

The energy necessary to propagate the crack within the model described above can be obtained using the classic Lake-Thomas [65] model of elastomer failure. Within this model $G_c = \Sigma n U$ where n is the number of main-chain bonds between crosslink or entanglement points, and U is the energy needed to break a main-chain bond. This relationship predicts a fracture energy of ~6 J/m^2 when Σ is 0.075 chains/nm^2, in reasonable agreement with the experimental results.

4
Optimum Toughening of the Interface: the Limits of G_c

Experimentally, the maximum fracture toughness that can be obtained for a given connecting chain and a given homopolymer pair is controlled by the maximum achievable value of the fibril failure stress σ_{fibril}.

Intuitively, the maximum value of σ_{fibril} should go through a maximum as a function of N since it is really the product of a force f to break or disentangle the chain (f should increase with N and then reach a plateau equal to f_b) and an areal density of connecting chains Σ_{sat}. This saturation areal density Σ_{sat} should decrease with the degree of polymerization N, for steric reasons. For a symmetric diblock copolymer, for example, that would form, if pure, a lamellar structure, Σ_{sat} should be essentially the same as the Σ of the lamellae and this Σ should decrease as $N^{-0.33}$ [36, 60]. Reactivity arguments given below impose at least as strong an N dependence on Σ_{sat} of copolymers formed at interfaces by reaction. Thus an optimum value of N should exist and this should correspond roughly to a value of N just where f saturates at f_b. Further increases in N will decrease Σ_{sat} without a compensating increase in f, leading to a decrease in σ_{fibril} with N. Interestingly, enough such arguments imply that the failure mechanism of the craze corresponding to this optimum N should involve some chain disentanglement. (If all the chains in the fibril fail by chain scission, $f = f_b$.)

Experiments support these theoretical arguments. The maximum attainable value of G_c as a function of N/N_e is shown in Fig. 32 for two diblock copolymer

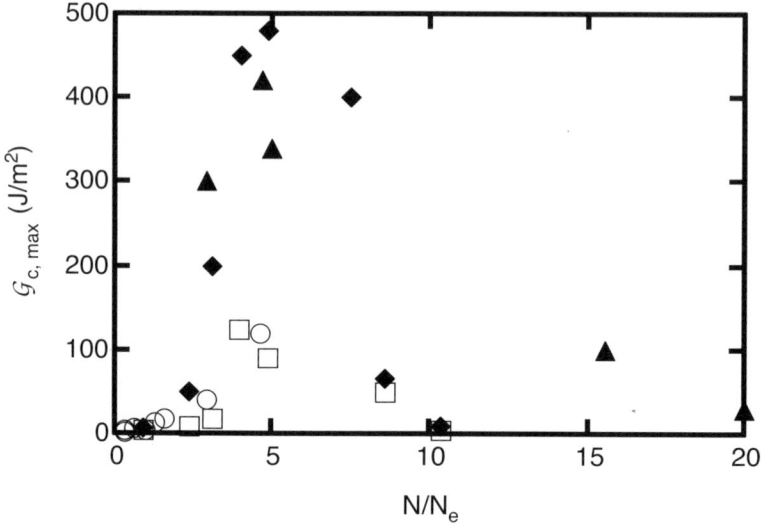

Fig. 32. Maximum achievable fracture toughness of interfaces between A and B polymers reinforced with block copolymers or end-grafted chains as a function of the degree of polymerization N of the reinforcing block. (▲) PS-b-PMMA between PPO and PMMA; (◆) dPS-COOH chains in a HIPS matrix grafted on an epoxy interface; (□) dPS-COOH chains in a PS matrix grafted at an epoxy interface; (○) PS-b-PVP chains at the interface between PS and PVP. Data from [22, 36, 38, 40]

systems (PS-PVP between PS and PVP and PS-PMMA between PMMA and PPO) where N is the degree of polymerization of the block, as well as for two grafted brush systems (dPS-COOH between epoxy and PS and dPS-COOH between epoxy and HIPS) where N is the degree of polymerization of the grafted chain. For both cases, N_e represents the average degree of polymerization between entanglements of the corresponding homopolymer.

For both the PS-PVP and the dPS-COOH epoxy systems the fact that a substantial fraction of dPS is found on the non-PS side of the interface after fracture at the maximum G_c shows that at least some chains in the craze fibrils fail by disentanglement, i.e. $f < f_b$.

In the case of the segregation of block copolymers at the interface, a quantitative prediction of Σ_{sat} is complicated by two main factors: unlike the case of grafted chains, one can obtain a higher apparent density of chains at the interface (by forming micelles or multiple lamellae) without actually increasing the areal density of connecting chains [33]; secondly, the experimental procedure, where the block copolymer is deposited at the interface and then annealed to allow local equilibration, might give rise to supersaturated interfaces, which are locally in equilibrium but not, strictly speaking, in equilibrium with the surrounding bulk homopolymers due to extremely slow kinetics.

An interesting issue in the case of reinforcement from block copolymers is the organization of the block copolymer at the interface when the areal density of copolymer exceeds that required to form a single brush. In this case, even though Σ of the original interface becomes saturated, the excess copolymer can lead to the formation of either additional weaker interfaces that become the new locus of fracture giving a decrease in G_c or to a roughening of the outer interface with the least craze-resistant homopolymer, leading to an increase in G_c.

While a unifying picture, encompassing all three types of copolymers, can be drawn for the effect of the connecting chain at low areal densities, this is no longer true at higher areal densities where the organization of the copolymer layer at the interface has very different mechanical effects whether it is a diblock or a triblock copolymer.

4.1
Diblock Copolymers

The situation for diblock copolymers has been examined in detail by Washiyama et al. in the PS-PVP system [33]. As shown in Fig. 33, they found that for symmetric copolymers of 510-540, the fracture toughness passed through a maximum at $\Sigma \approx 0.2$ chains/nm^2 corresponding to a nominal areal density of chains a little higher than what is contained in a pure block copolymer lamella of thickness L. At higher values of Σ, TEM observations showed that a multilamellar structure formed at the interface, as illustrated in Fig. 34 for the 510-540 copolymer. The fracture toughness of such an interface dropped sharply to stabilize then at a plateau value. The areal density at which G_c reached this plateau value corresponded closely to what would be expected for the thickness of 3/2 L. The

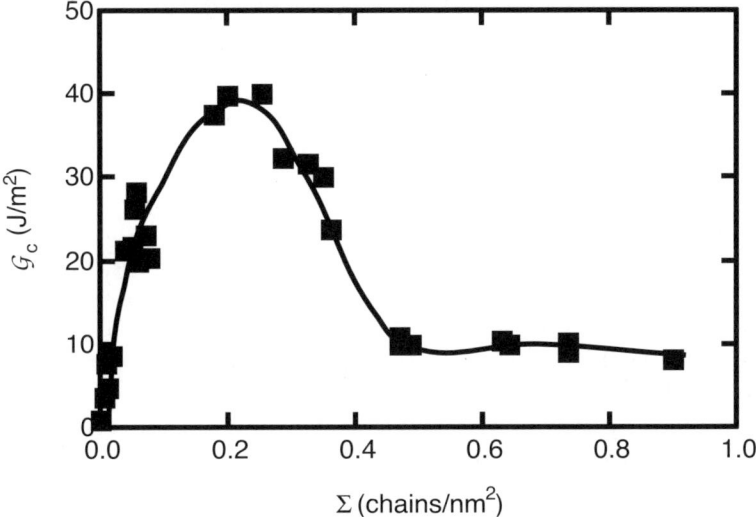

Fig. 33. G_c vs. Σ of interfaces between PS and PVP reinforced by 510-540 dPS-b-PVP diblock copolymers [33]

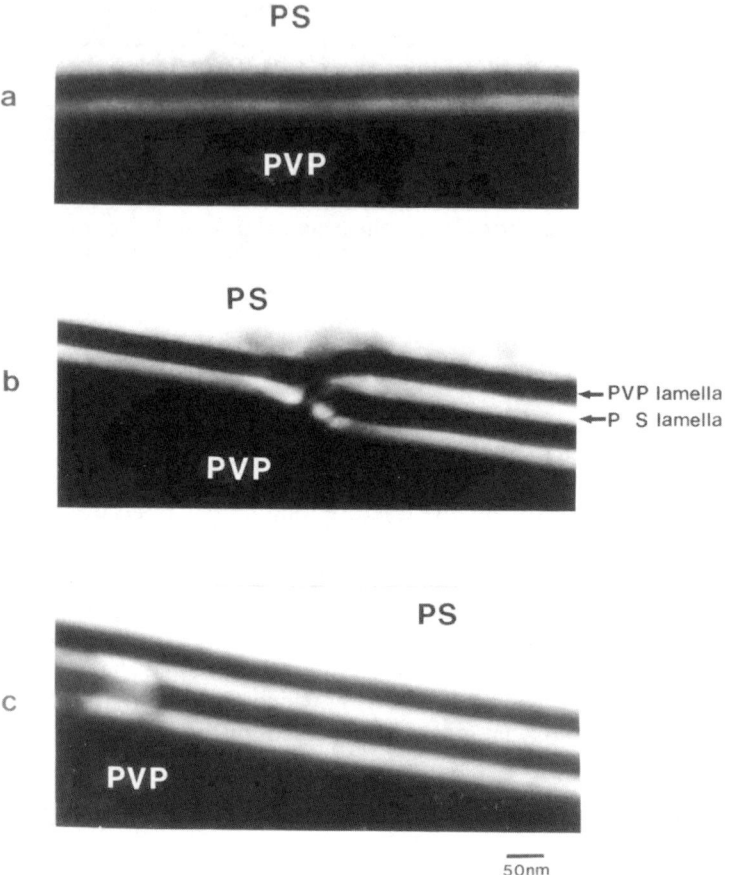

Fig. 34. TEM micrographs [33] of the undeformed interface between PS and PVP reinforced with a 510-540 dPS-b-PVP diblock copolymer. **a** $\Sigma = 0.4$ chains/nm^2; **b** $\Sigma = 0.7$ chains/nm^2; **c** $\Sigma = 0.85$ chains/nm^2

analysis of the fracture surfaces in Fig. 35a showed that in that high Σ regime the crack path goes through the weaker PS lamella which is formed by the contact of two block copolymer blocks, as shown schematically in Fig. 35b for the 510-540 copolymer. The dramatic weakening of the interface was attributed to the swelling of this PS lamella by the low molecular weight component of the polydisperse homopolymer PS. This low molecular weight component greatly reduced the craze stability by favoring disentanglement in the craze fibrils. A comparison between TEM micrographs of the lamellae and the surface analysis of the fracture surfaces with FRES confirmed this hypothesis.

It should be noted, however, that all the decrease in fracture toughness cannot be attributed to the presence of the low molecular weight fraction. Recent experiments by Dai et al., on the 510-540 copolymer with a monodisperse high molec-

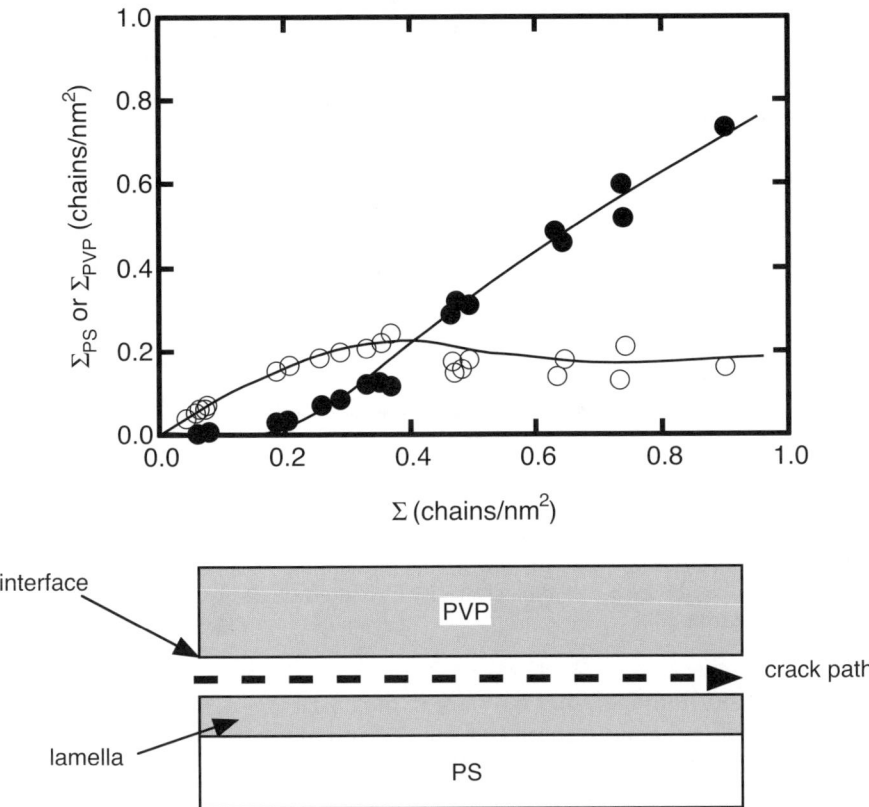

Fig. 35. a Surface analysis after fracture of the interface between PS and PVP reinforced with a 510-540 dPS-b-PVP diblock copolymer showing the fraction of deuterium on the PS side (●) and on the PVP side (○). **b** Schematic of the crack path in the block copolymer lamella. Data from [33]

ular weight PS homopolymer, also showed a decrease in G_c after the maximum, albeit of smaller amplitude [66]. This result reflects the fact that, for the 510-540 homopolymer, the PS block/PS block interface is intrinsically weaker than the PS block/PS homopolymer interface. One would expect this effect to be smaller or to disappear for longer diblocks.

A similar result was obtained by Creton et al. on PS-PMMA between PMMA and PPO [37]. In their case as well, SIMS surface analysis showed that the maximum in G_c was obtained for a value of Σ slightly higher than $L/2$. Similarly to the PS-PVP case, at higher values of Σ, the measured G_c decreased and reached a plateau. SIMS results were consistent with the formation of multiple lamellae with the crack always going through the PS lamella closer to the PMMA side of the interface. Creton et al. compared the fracture toughness of the interface in the high Σ regime, where it is controlled by the PS lamella, to the value of G_c obtained for

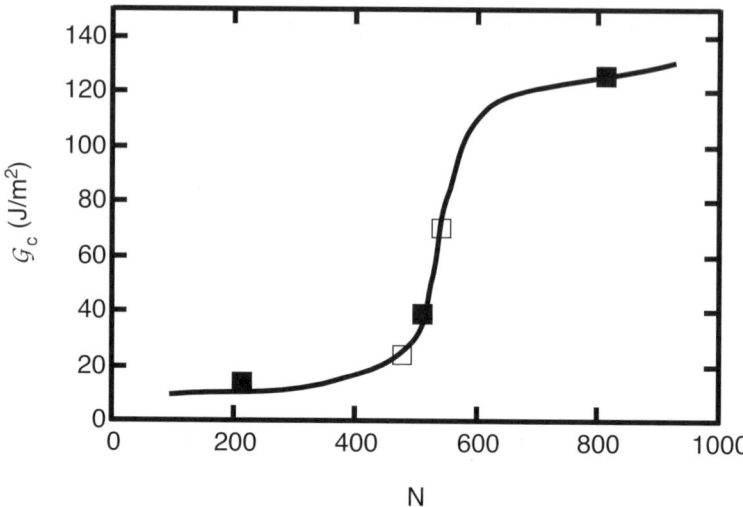

Fig. 36. Comparison of the G_c values at multilayer coverage for the PS-b-PMMA diblock at a PPO/PMMA interface (□), where failure occurs through the PS lamella, with the G_c maximum observed for the PS-*b*-PVP diblock at a PS/PVP interface (■), where failure occurs in the brush. Data from [37]

the PS-brush/PS homopolymer interface in the PS-PVP system. All these results can be plotted on the same graph (see Fig. 36) and imply that for the PS-PMMA between PMMA and PPO, the PPO, if it swells the PS lamella, does not alter significantly its resistance to fracture.

On the other hand, for strongly asymmetric diblocks such as the PS-PVP 580-220, a similar decrease of G_c with Σ is not observed. In this case, TEM observations of the interfacial area show that the excess diblock forms spherical micelles that do not affect the fracture toughness of the interface [33].

4.2
A-B-A Triblock Copolymers

At very high values of Σ, greatly exceeding the theoretical value for single layer coverage, triblock copolymers have a strikingly different behavior to diblocks. This behavior is illustrated in Fig. 37 that compares the effect of the 580-1620-580 PVP-dPS-PVP triblock with that of the 800-870 diblock at high values of Σ. G_c does not go through a maximum but rather continues to increase up to a coverage of $\Sigma = 0.4$ chains/nm². The maximum observed value of G_c greatly exceeds the maximum G_c that could be obtained with the 800-870 diblock and approaches the G_c of bulk PS. A tentative explanation for this peculiar behavior is suggested by TEM observations of microtomed sections near the interface. These micrographs show that, at high coverage, the 580-1620-580 triblock formed a microphase separated structure (Fig. 38) formed of cylinders oriented perpendic-

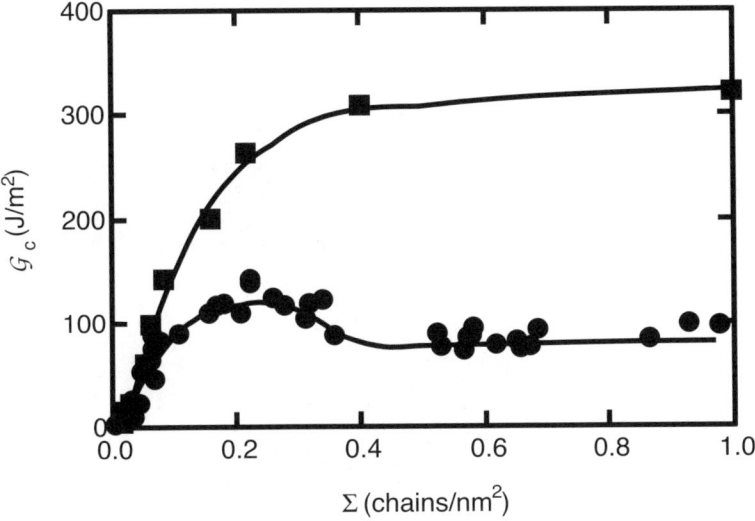

Fig. 37. Comparison of the fracture toughness of interfaces between PS and PVP reinforced with a 800-870 dPS-PVP diblock (●) and with a 580-1620-580 PVP-b-dPS-b-PVP triblock (■). While G_c is quite similar for both block copolymers at low coverage, the triblock is much more effective at higher levels of coverage as explained in the text. Data from [33, 46]

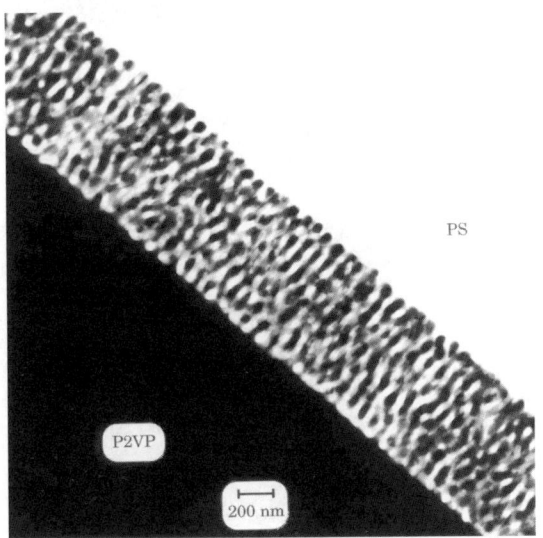

Fig. 38. TEM micrograph [46] of an undeformed interface between PS and PVP reinforced with a 580-1620-580 triblock copolymer. $\Sigma > 0.1$ chains/nm^2

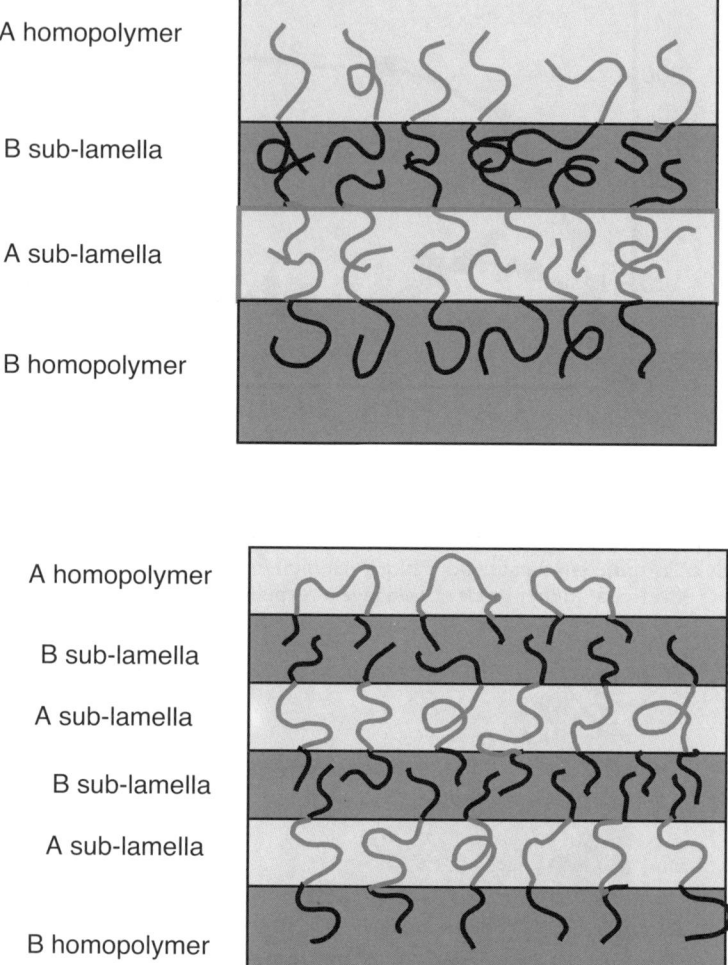

Fig. 39. Schematic of the conformation of block copolymers inside lamellae: **a** diblocks and **b** triblocks. The bridging effect of triblock copolymers is responsible for their better mechanical reinforcement at high copolymer coverage

ular to the interface and a significant roughening of the interface was observed. This very rough interface with no obvious weak points may be responsible for crack deviations in the bulk PS or for an additional nucleation of 45° crazes at the crack tip that would have increased G_c.

A similar irregular phase-separated structure (albeit lamellar this time) was also observed for the 290-470-290 triblock. In this case, since the lamellae are oriented parallel to the interface, the stronger mechanical reinforcement is probably due to the bridging effect of triblock chains, as schematically shown in Fig. 39.

5
Interfaces Between Polymers, Coupling by Random Copolymers

While adding connecting chains is an effective way to reinforce polymer interfaces, there are situations where that solution may be impractical. An alternative option is to increase the average degree of interpenetration of the two immiscible polymers, effectively broadening the interfacial width. In this case, however, there is no longer a well-defined areal density of chains Σ or a well-defined interpenetration degree of polymerization N, but rather a distribution of N's.

The microscopic analysis of the fracture mechanisms developed in Sect. 4 is expected to hold for these interfaces as well except for the fact that σ_{fibril} is no longer given by Eq. (25) but is, in the general case, a complicated function of the molecular weight of the polymers and of the χ interaction parameter. For molecular weights much larger than M_e, σ_{fibril} should be a function of the interfacial width only. Assuming that the molecular weight of the polymers cannot be varied much, this means controlling the value of the interaction parameter χ between the two materials.

5.1
Interfaces Between Homopolymers

From the point of view of the mechanical reinforcement it is important to understand how the mechanical strength of the interface varies when the χ parameter changes (if the two polymers are immiscible) and, more generally, how the fracture toughness of the interface depends on the interfacial width a_I of the interface [as defined in Eq. (3)].

The development of neutron reflectivity as a tool to investigate polymer interfaces has now made it possible to measure G_c and a_I on the same polymer interfaces, and the results of Schnell et al. [67, 68] on interfaces between styrenic polymers are shown in Fig. 40a. These results were obtained for three different types of interfaces:
1. Interfaces between monodisperse high molecular weight PS. In this case a healing experiment was performed and a_I and G_c were measured as a function of annealing time.
2. Interfaces between monodisperse high molecular weight PS and monodisperse high molecular weight poly(paramethylstyrene) of various molecular weights. Since the χ parameter of this polymer pair is of the order of 0.003–0.005, the range of interfacial widths which could be obtained by varying the molecular weight of the polymers between 150 kg/mole and 1250 kg/mole and the annealing temperature from 140 to 185 °C, was 9–13 nm.

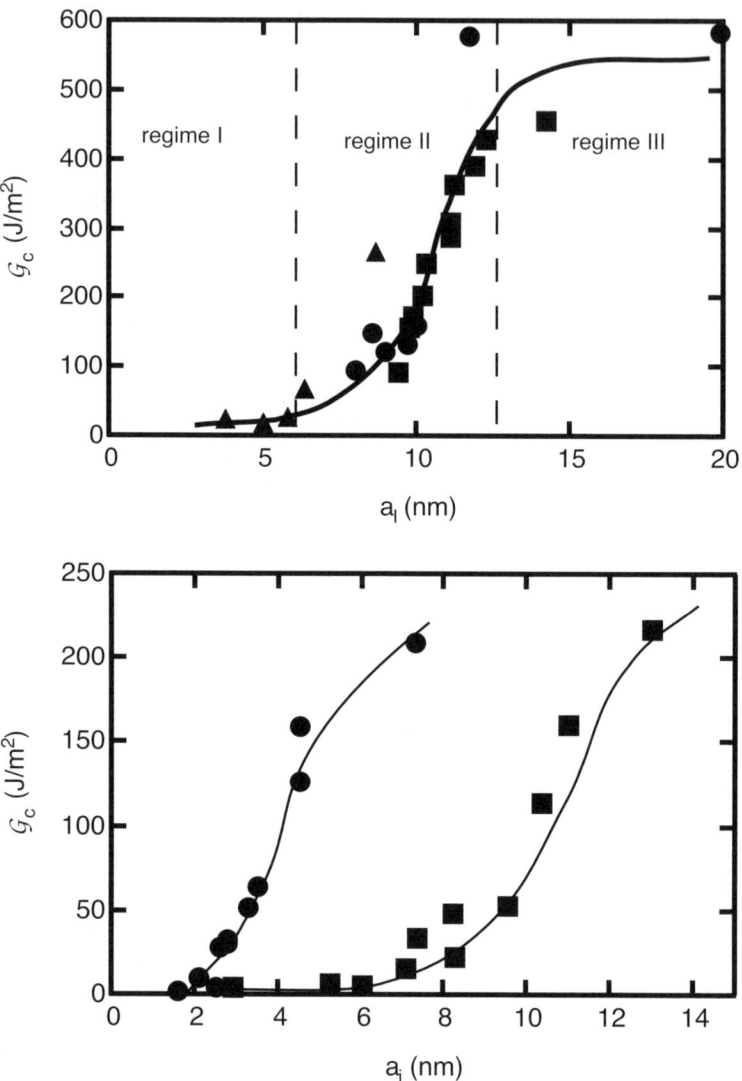

Fig. 40. Fracture toughness G_c of interfaces between glassy homopolymers as a function of their width a_I. **a** (●) PS/PS interfaces; (■) PS/PPMS interfaces; (▲) PS/PBrxS interfaces. **b** (●) PS-r-PVP/PS interfaces, (■) PS-r-PMMA/PS and PS-r-PMMA/PMMA interfaces. Data from [67, 69, 70]

3. Interfaces between high molecular weight monodisperse PS and monodisperse high molecular weight brominated PS (PBr_xS) where x was varied between 0.11 and 0.33. In this case χ varied between 0.1 and 0.001 depending on the degree of bromination and a different range of interfacial widths (narrower) was obtained than in case 2.

The results of Fig. 40a clearly show three different regimes which can be interpreted in terms of microscopic failure mechanisms.
- For $a_I < 6$ nm, G_c is low and presumably the failure mechanism is simple chain pullout or simple chain scission with no plastic zone being formed ahead of the crack tip.
- For 6 nm $< a_I <$ 12 nm, G_c increases sharply with interfacial width, suggesting a transition from failure by simple fracture or pullout, to failure after formation of a craze at the crack tip. This craze is clearly observable in the PS-PMMA studies. Once the transition is passed, the increase in G_c is due to the increase in the maximum stress σ_{fibril} that can be sustained by the fibrils since all polymers in this study have very similar crazing stresses.
- Finally, for $a_I > 12$ nm, the fibril failure stress becomes independent of the interfacial width and G_c is that of the bulk homopolymer.

The unique dependence of G_c on the width of the interface has also been found for interfaces between high molecular weight random copolymers and homopolymers [69, 70]. It is apparent from Fig. 40b, however, that while regimes I and II are also observed, the critical width, where the transition from regime I to regime II occurs, varies for different PS-based systems even though the bulk entanglement structure is the same.

These three regimes can be visualized in Fig. 41. This figure shows the fibril failure stress as a function of a_I as well as the value of fibril failure stress one would expect for bulk PS based on the value of strands crossing the interface in

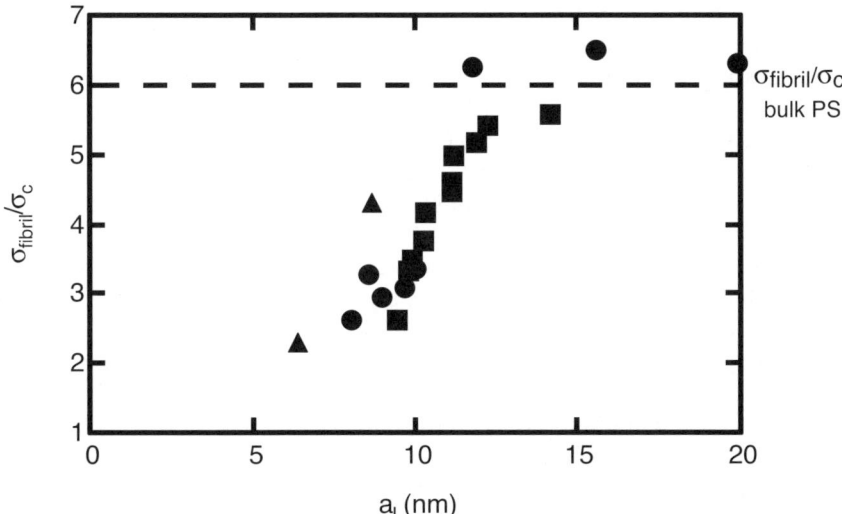

Fig. 41. Normalized fibril failure stress σ_{fibril}/σ_c as a function of the width of the interface a_I for (●) PS/PS interfaces; (■) PS/PPMS interfaces; (▲) PS/PBrxS interfaces. The *dashed line* represents the fibril failure stress one would expect from bulk PS

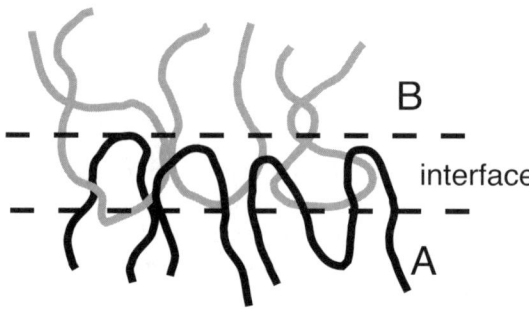

Fig. 42. Schematic of the loop structure of the chains at the interface that the fracture toughness data suggests

bulk PS. Two interesting results can be extracted from this study: The transition from chain pullout to crazing can be interpreted as the point where the stress that can be sustained by the interpenetrated chains (the conformation of which is not precisely known) is higher than the crazing stress.

Schnell at al. argue that the low degree of interpenetration necessary to activate the plastic deformation mechanisms could be indicative of the formation of a significant number of loops at the interface rather than chain ends, as shown schematically in Fig. 42. This argument would apply even more strongly for the results on PS-r-PVP/PS interfaces recently reported by Benkoski et al. [70] but would not apply to PMMA interfaces where the transition from regimes I and II occurs for much wider interfaces [69].

The transition from regime II to regime III is indicative of the fact that the interface can no longer be distinguished from the bulk by its fracture mechanism. It is interesting to point out that for these high molecular weight PS and PMMA samples, the a_I at which this transition occurs is much closer to the rms end-to-end distance between entanglements (rms end-to-end distance of a chain of molecular weight M_e) in the bulk than to the radius of gyration of the chains.

Brown has proposed a simple model to calculate the amount of coupling by entangled loops and found that, in the PMMA/random copolymer system, the model predicted a higher toughness than that observed for narrow interfaces. He suggested that this deviation could be caused by a decrease in entanglement density close to narrow interfaces.

In principle these concepts should be generally applicable to any polymer pair, so that one would be tempted to argue that a well-entangled polymer such as polycarbonate should need only a very short interpenetration length to give a strong interface with itself. However, one must keep in mind that the crazing stress is likely to increase with the decrease in average molecular weight between entanglements, effectively offsetting part of the advantage of having a smaller M_e. Furthermore, based on the results shown in Fig. 40, one cannot exclude that two similar entanglement structures at the interface could have different abilities to transfer stress.

5.2
Random Copolymers at Interfaces Between Homopolymers

While using block copolymers is an effective method to increase the mechanical strength of an interface between two immiscible polymers, it is not the only one. Early experiments of Char and Brown [36] showed that random copolymers could also be effective reinforcing agents. They used PS-PMMA random copolymers containing 70% styrene at interfaces between PS and PMMA, and PPO and PMMA. The maximum toughness obtained was ≈90 J/m^2 for the PS/PMMA interface, almost a factor of 20 higher than the toughness of the bare interface.

Following these early studies, several groups have investigated more systematically the relationship between the molecular composition f of the random copolymer A_f-r-$B_{(1-f)}$, the respective χ parameters between the random copolymer and the homopolymers on either side of the interface, and the fracture toughness of the interface G_c. [26, 71–78]. In parallel to these experimental studies, theoretical arguments have been developed to account for the mechanical reinforcement provided by the random copolymers.

While early interpretations of the experimental results focused on a picture of the random copolymer weaving back and forth on both sides of the interface [79, 80], as shown in Fig. 43a, Milner and Fredrickson [81] later pointed out that long random copolymer chains composed of monomers with a large interaction parameter should be immiscible with their respective homopolymers and therefore should collapse at the interface forming either: (1) a 2D phase-separated structure at low values of Σ, as shown in Fig. 43b, or (2) a separate layer of A-r-B sandwiched between polymer A and polymer B at high values of Σ, as shown in Fig. 43c.

While the reinforcement effect at low areal density is still poorly understood due to the occurrence of this 2D phase separation, at high coverage, the problem can be reduced to the replacement of a narrow interface by two wider ones. The width a_I of an interface between polymers A and B is given by Eq. (3), where χ_{A-B} is the interaction parameter between monomers A and B. The question is then to define a reasonable value of the interaction parameter between the random copolymer and the two polymers on either side of the interface. In the simplest case, where we are considering a A_f-r-$B_{(1-f)}$ random copolymer at the interface between homopolymers A and B, these equivalent interaction parameters can be written as:

$$\chi_{rcp-A} \approx (1-f)^2 \chi_{A-B} \tag{27}$$

$$\chi_{rcp-B} \approx f^2 \chi_{A-B} \tag{28}$$

In the presence of a thick enough random copolymer layer, the system is now composed of two interfaces in series so that it is the weaker of the two interfaces (i.e. the narrowest) which will control the failure process. The optimum toughening is then expected to occur for $f = 1/2$ in this case, i.e. a symmetric broadening of the interface.

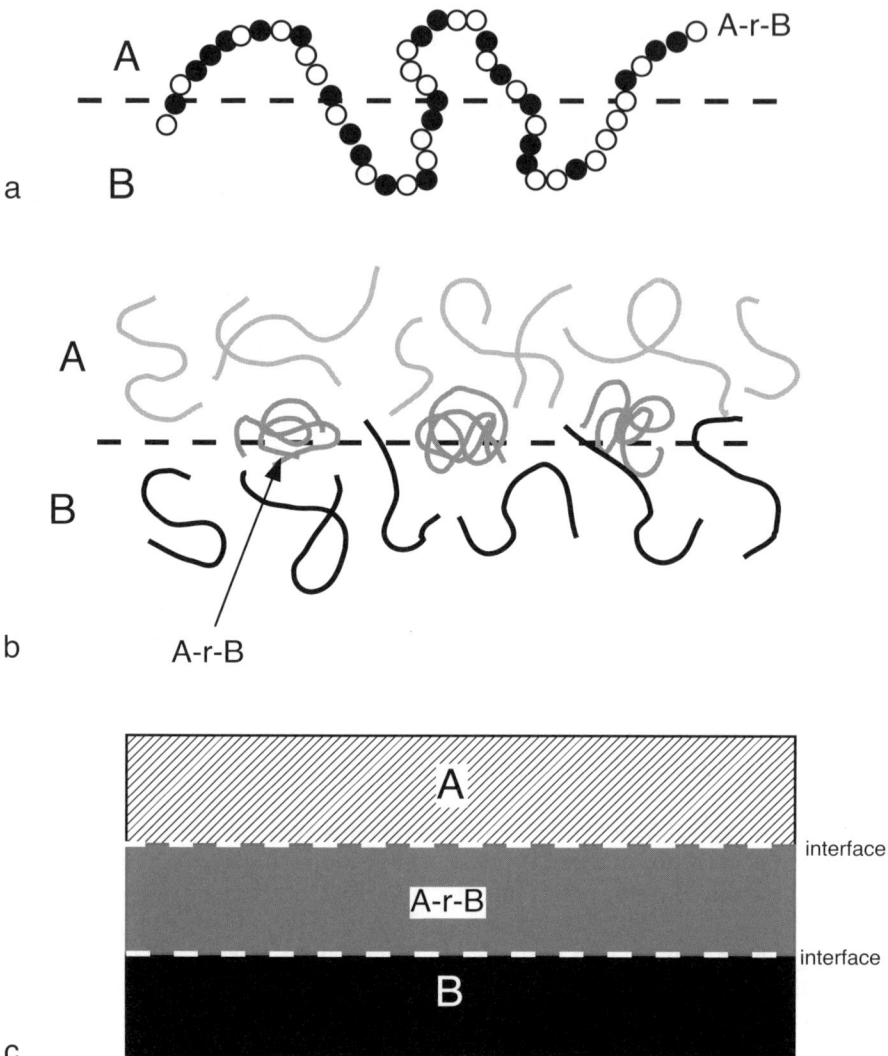

Fig. 43. Schematic of the conformation of A-r-B random copolymer chains at the interface between A and B homopolymers. **a** A single chain; **b** at low coverage (low Σ); and **c** at high coverage (high Σ)

This is indeed the case for the reinforcement of a PS/PVP interface with a series of dPS$_f$-r-PVP$_{(1-f)}$ random copolymers with variable values of f [59, 75, 79]. The value of the fracture toughness G_c obtained for thick copolymer layers is shown in Fig. 44 as a function of the average composition f. It clearly shows a sharp maximum for a value of f close to 0.5, as predicted by the weakest link argument presented above. The situation becomes slightly more complicated how-

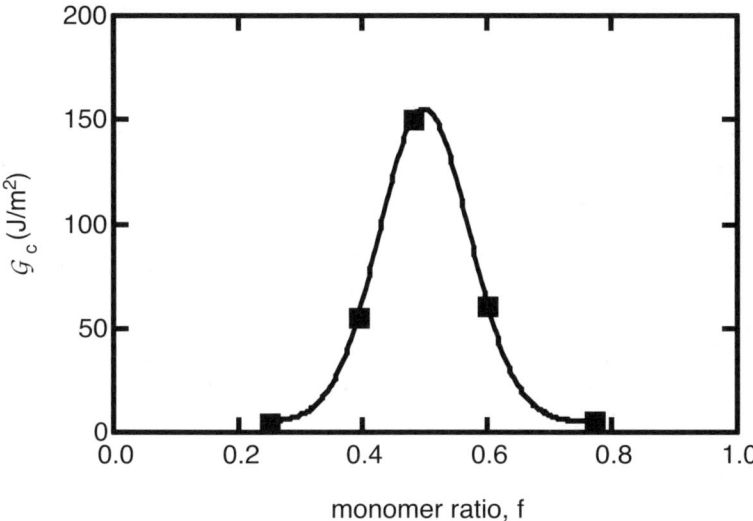

Fig. 44. Fracture toughness of interfaces between PS and PVP reinforced by a thick layer of a PS_f-r-PVP^{1-f} random copolymer. The pronounced maximum as a function of f reflects the situation where $\chi_{rcp\text{-}PS} = \chi_{rcp\text{-}PVP}$. Data from [75]

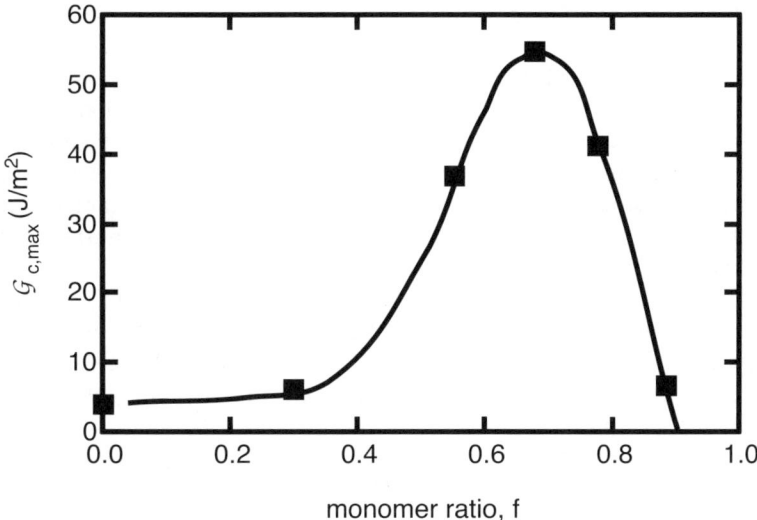

Fig. 45. $\mathcal{G}_c$ as a function of monomer ratio f for thick layers of PS_f-r-$PMMA_{1-f}$ random copolymers at the interface between PS and PMMA. Clearly the optimum reinforcement is not for $f = 0.5$ and reflects the composition dependent χ between PS and PMMA. Data from [72]

ever when PS$_f$-r-PMMA$_{(1-f)}$ is used to reinforce interfaces between PS and PMMA. In this case the maximum in G_c is observed for 68% styrene content in the random copolymer, as shown in Fig. 45 [72].

Measurements of the interfacial width of the interfaces by neutron reflectivity [73] show that this composition corresponds to the situation where the interface is symmetrically broadened. However, the simple approximation used to obtain $\chi_{rcp\text{-}PS}$ and $\chi_{rcp\text{-}PMMA}$ no longer holds since $\chi_{PS\text{-}PMMA}$ is composition-dependent in this case.

On a more careful analysis of the PS-PVP results, a further discrepancy between theory and experiment appears. While Eqs. (27) and (28) correctly predict the location of the maximum in G_c, the value of $\chi_{rcp\text{-}PS}$ for $f = 0.5$ should be about 0.025 based on a value of $\chi_{PVP\text{-}PS}$ of 0.1 [82]. An interface between two homopolymers with such a value of χ would have an interfacial width of approximately 30 Å and be very weak mechanically [67].

This discrepancy is partly due to the fact that random copolymers produced by a batch free-radical polymerization synthetic method can have a significant composition drift if the respective reactivity ratios of the monomers are different.[2] This means that the value of the parameter f is not homogeneous in the copolymer layer at the interface. In the PS-PVP case discussed above, the random copolymers directly in contact with the PS or the PVP side of the interface at equilibrium would be PS-rich or PVP-rich, respectively. This segregation of copolymer fractions to their preferred interfaces gives rise to a broadening of these interfaces relative to the case of a random copolymer with a narrow distribution of f values.

This effect has been clearly demonstrated by reinforcing a PS/PVP interface with a thick layer of a 50:50 blend of PS$_{0.77}$-r-PVP$_{0.23}$ and PS$_{0.25}$-r-PVP$_{0.75}$. The fracture toughness of such an interface was found to be ~100 J/m² while each one of these copolymers if used separately at the interface gave $G_c < 5$ J/m². A similar graded layer is probably formed when a high molecular weight PS-PMMA diblock copolymer is used as a coupling agent between PS and PMMA. When the annealing time is short, as χ is small, the copolymer does not organize at the interface, which is much tougher than expected for the organized diblock. With extended high temperature annealing the copolymer organizes at the interface and the toughness drops [34].

Another situation which has been investigated in some detail is that of an A$_f$-r-C$_{(1-f)}$ copolymer at the interface between A and B [76–78]. In this case the two effective χ parameters $\chi_{rcp\text{-}A}$ and $\chi_{rcp\text{-}B}$ are no longer dependent only on f and $\chi_{A\text{-}B}$ but are very much determined by the respective interactions of the C monomer with A and B:

$$\chi_{rcp\text{-}A-} = f^2 \chi_{C\text{-}A} \text{ and } \chi_{rcp\text{-}B} = (1-f)\chi_{A\text{-}B} + f\chi_{C\text{-}B} - f(1-f)\chi_{C\text{-}A} \qquad (29)$$

---

2 The hypothesis that a PS-r-PVP copolymer produced in a way that precludes composition drift would produce lower value of G_c (52.5 J/m² vs. 150 J/m² for $f = 0.5$) is confirmed by recent measurements [70]; the data in Fig 40b are corrected for composition drift.

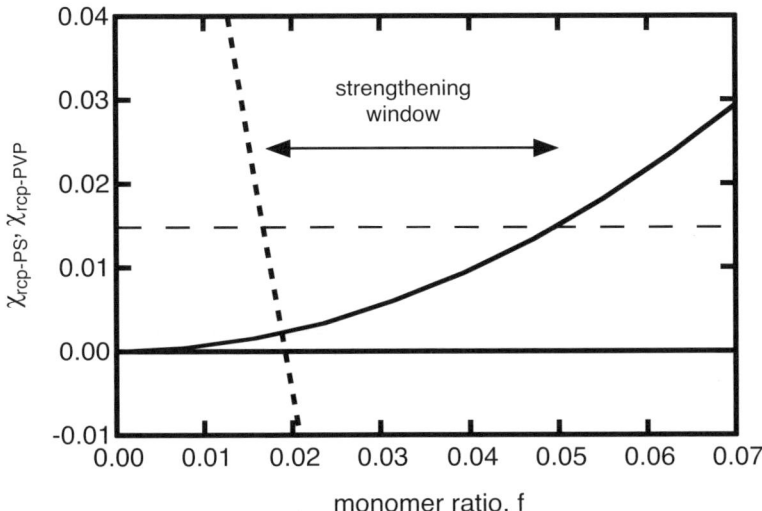

Fig. 46. Effective interaction parameters $\chi_{rcp\text{-}PS}$ (—) and $\chi_{rcp\text{-}PVP}$ (- -) as functions of monomer ratio in the copolymer f for a poly(styrene-d8-co-4-hydroxystyrene) where f is the molar fraction of 4-hydroxystyrene. The *dotted line* ($\chi = 0.015$) is the maximum value for which a strong rcp/homopolymer interface is formed

By using these equations and with prior knowledge of the three monomer interaction parameters, it is possible to predict the composition f giving rise to the optimum broadening of both interfaces and therefore to the maximum in fracture toughness G_c, as schematically shown in Fig. 46. Conversely, if some of the interaction parameters are not known, the optimum value of f can provide information on the missing χ values.

In the specific case where A and B are PS and PVP, respectively, and C is poly(styrene-r-4-hydroxystyrene), poly(styrene-r-4-vinylbenzamide) or poly(styrene-r-4-vinyl-N-ethylbenzamide), the maximum of G_c as a function of f and the locus of fracture are well predicted by Eq. (29) by using values of χ parameters obtained independently showing conclusively the effect of different A-C and B-C interaction strengths. Interestingly, the strong rcp/PVP interface is not due to a large attractive χ between PVP and the polar monomer but rather to the very large repulsive χ between the PS and the polar monomer which tends to expel it from the PS-rich-copolymer layer. This is a manifestation of the so-called copolymer effect [83, 84] for inducing miscibility in polymer blends. It also appears that the value of χ at the symmetric broadening point (where $\chi_{rcp\text{-}PS} = \chi_{rcp\text{-}PVP}$) increases with decreasing repulsive χ between the polar C unit and the PS consistent with the increase in maximum G_c.

Surface analysis of the fracture surfaces also show conclusively that the weak interface is always the rcp/PS interface for values of f above the optimum and the rcp/PVP for values of f below the optimum, as illustrated in Fig. 47.

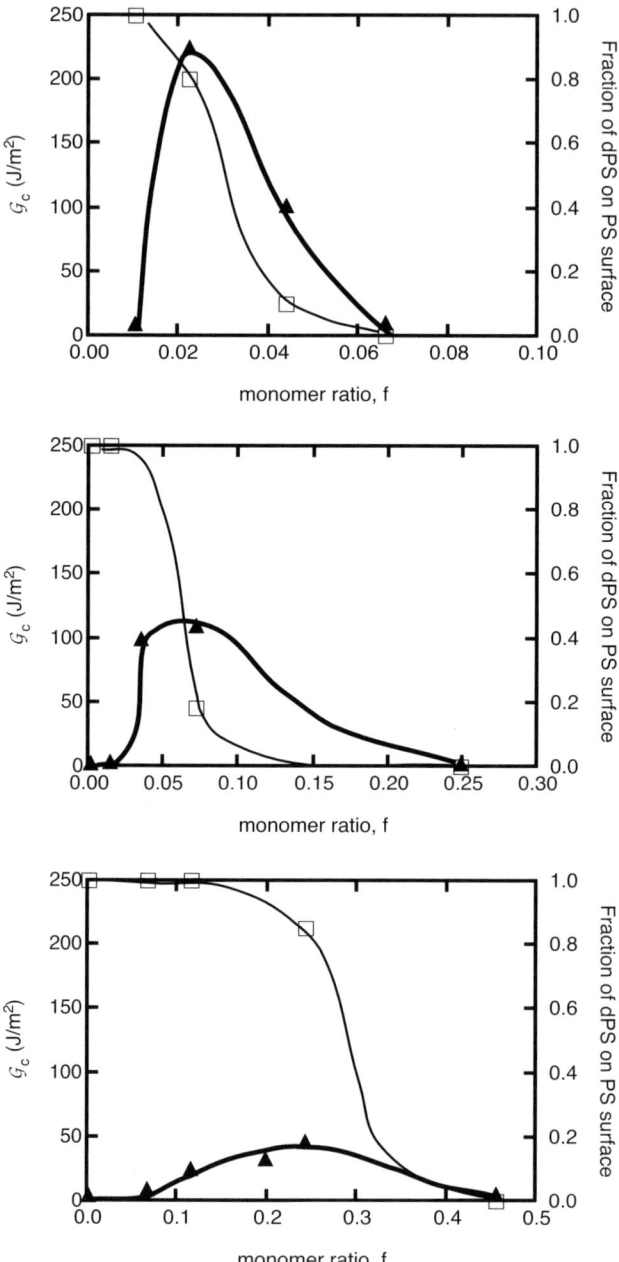

Fig. 47. Fracture toughness $\mathcal{G}_c$ (▲) at high copolymer coverage (saturation values), and fraction of deuterium on the PS side of the interface after fracture (□) at low copolymer coverage, as a function of copolymer composition f. **a** Poly(styrene-d8-co-4-hydroxystyrene); **b** poly(styrene-d8-co-4-vinyl-N-ethylbenzamide); and **c** poly(styrene-d8-co-4-vinylbenzamide). Data from [77]

Since poly(hydroxystyrene) has the ability to hydrogen bond with PVP, one could have attributed the reinforcement to specific interactions between PS(OH) and PVP groups. However, this interpretation is not consistent with the low toughness observed for low values of f. At long enough annealing times, a high value of G_c due to the progressive diffusion of PS(OH) groups to the PVP interface should have been observed if hydrogen bonding was important.

6
Reactive Systems

After considering systems where preformed copolymers are *added* at the interface, a logical extension is to consider reactive systems where copolymers are formed in situ at the interface. This method, which is widely used industrially, is schematically described in Fig. 48. Several procedures can be used to achieve the same goal but the general strategy is the following. A small amount of polymer, functionalized with a suitable reactive group, is blended in one or both of the homopolymers. The reactive group is usually incorporated either as a comonomer or as an end-group.

The reactive groups on both sides of the interface are chosen in such a way that once they come into contact at the interface, they are able to mutually react, form a strong chemical bond and remain pinned to the interface. The major drawback of this method is that it is difficult to produce copolymers at the interface that are both long enough to be well entangled with the homopolymer phase on either side and dense enough (high Σ) to prevent failure by chain scission.

From the theoretical viewpoint the problem is rather complicated as it requires a good understanding of both surface chemistry and polymer physics [39, 85–88]. While a detailed discussion of the interfacial reaction kinetics is beyond the scope of this review, it is useful to outline the main stages of formation of the interfacial copolymer layer. We will follow a published report [86] and consider the common case of a dilute concentration of chains carrying reactive groups on both sides of the interface, as schematically described in Fig. 49. At early stages of the copolymer layer buildup, one expects the concentration of reactive groups in the bulk near the interface to remain relatively constant so that the reaction kinetics is controlled by the interfacial reaction rate. Hence, initially, the areal density of chains formed, Σ, is expected to increase linearly with time. If the reaction rate is fast enough, the characteristic time for this first process (given roughly by the ratio of the diffusion coefficient over the square of the interfacial rate coefficient) controls the concentration of reactive groups near the interface which decrease significantly. If the reaction rate is slow, the copolymer areal density becomes sufficiently high to form a chemical potential barrier of the order of several $k_B T$ for a new chain to graft and the areal density of copolymer chains saturates before this characteristic time for depletion is reached. Under these conditions diffusion never becomes the rate-limiting process. For fast interfacial reaction, a depletion hole can form near the interface in the intermediate stage

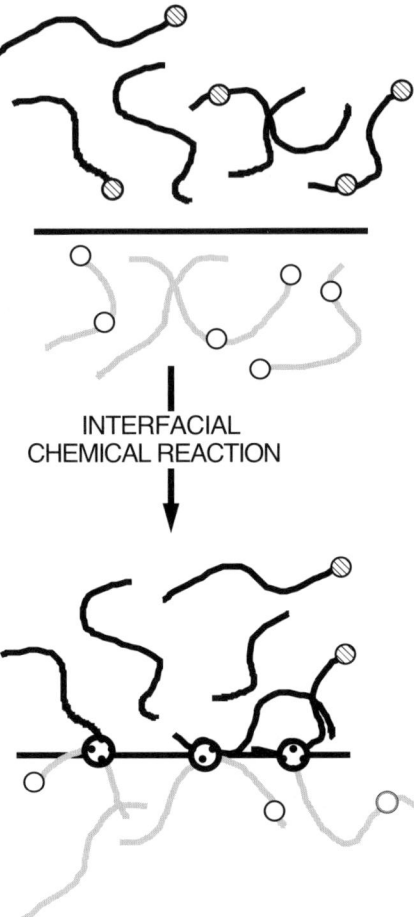

Fig. 48. Schematic of the interfacial reaction of reactive chains. Polymer chains on either side of the interface contain functional groups that can mutually react together, effectively forming a copolymer in situ

and the copolymer buildup will now be controlled by the diffusion coefficient of reactive species to the interface so that Σ increases with $t^{1/2}$. In the final stage, as for the slow reaction case, the areal density of copolymer chains saturates. This saturation regime is expected to occur when the coverage is of the order of:

$$\Sigma_{sat} \approx b^{-2} N^{-1/2} \tag{30}$$

where b is the statistical segment length. An increase in N leads not only to a decrease in the reaction kinetics at low Σ under conditions where diffusion is not the rate-controlling process [89], but also to a decreasing value for the areal chain density at saturation. This N dependence of Σ_{sat} and of the reaction kinetics makes it difficult to graft chains that are both sufficiently long enough to be

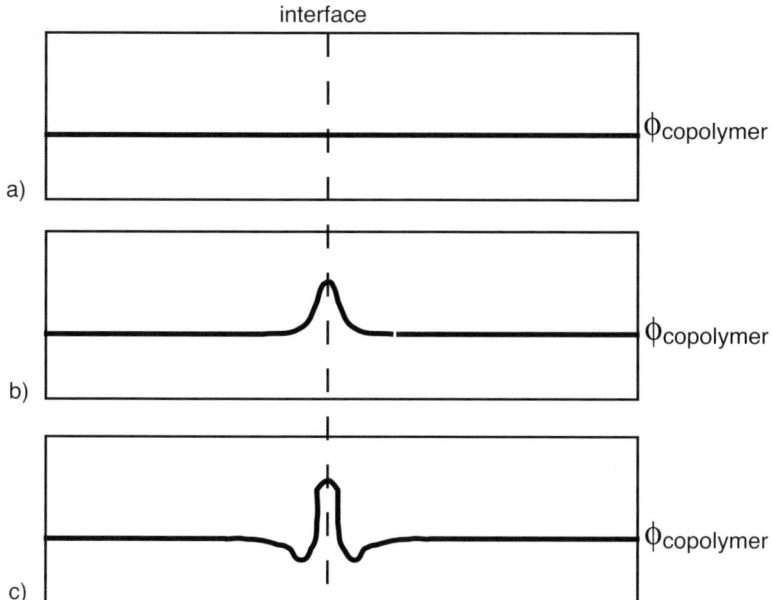

Fig. 49. Schematic of a dilute concentration ϕ of functionalized A and B chains diffusing and reacting at the interface. **a** Initial situation; **b** reaction controlled kinetics; and **c** diffusion controlled kinetics. Both situations can occur at different stages of the reaction process

entangled but also sufficiently dense to be above Σ^* so that crack propagation along the interface will involve crazing rather than only chain scission. It is worthwhile pointing out that the interface cannot be supersaturated with block copolymer in reactive systems since the existing chains prevent any further formation of copolymer. However, if the chains are short enough, the interfacial tension can be driven to zero before saturation occurs and an interfacial instability leading to the formation of a microemulsion occurs [90, 91].

Experimentally, also, reactive systems are much more complex than the preformed block copolymer systems because, typically, the reactive species have a distribution of molecular weights and may have several reactive groups per chain. Then there is not only the issue of how much of the graft copolymer is formed at the interface (Σ) but also of which species react, e.g. is there preferential reaction of low N species?

Since reactive systems have a great practical relevance, many attempts have been made in recent years to adapt the techniques used for pre-made block copolymers to obtain more detailed information on the correlation between the interfacial structure and fracture toughness. The simplest case is that of a dilute solution of monodisperse chains with one reactive group per chain, located at the chain end. In this case, the mechanical problem of reinforcement becomes similar to that of single connector diblock copolymers discussed earlier. The important issues are then mostly related to the competition between grafting kinetics and diffusion of the reactive groups to the interface. In the case of dPS-COOH chains reacting on an epoxy surface [38], Norton et al. found that the grafting rate at the interface was controlled by reaction, not diffusion, and that the maximum observed grafting density was strongly decreased as N of the dPS-COOH increased. Epoxy stoichiometry mattered as the grafting reaction proceeded about ten times as rapidly if excess epoxy groups were present in the network than if excess amine groups from the amine hardener were present [39].

In a different system with polydisperse reactive chains of maleic anhydride end-grafted polypropylene (PP-g-MA) reacting on a polyamide 6 (PA-6), Boucher et al. produced evidence consistent with a grafting rate controlled by the center-of-mass diffusion of PP-g-MA to the interface [92].

Other experimental studies have mostly focused on the relationship between areal density of connecting chains and G_c or, if Σ could not be determined, on the relationship between G_c and annealing time and temperature.

6.1
Reactive Polymers with Multiple Functional Groups

Although the telechelic functional polymers are very attractive from a fundamental point of view, their synthesis is often impossible. Much more commonly, the active groups are incorporated in the chain either by a free-radical copolymerization with a small amount of functionalized comonomer or by functionalization of the chain after polymerization in the presence of free radicals (typical of the functionalization of the polyolefins). Either method generally produces several reactive sites per chain.

In terms of mechanical reinforcement, the existence of multiple reactive groups raises some specific questions. When multiple reactive sites are present on the chain, an important question is the average number of those sites per chain that actually react and how that affects the mechanical properties of the interface.

Lee and Char [93] studied the reinforcement of the interface between an amorphous polyamide (PA) and polystyrene with the addition of thin layers of a random copolymer of styrene-maleic anhydride (with ~8% MA) sandwiched at the interface. After annealing above the T_g of PS, they found significantly higher values of G_c for samples prepared with thinner layers of SMA than for the thicker ones. They initially rationalized their results by invoking the competition between the reaction rate at the interface and the diffusion rate of the SMA away from the interface. For very thick layers, and therefore also for pure SMA, the reaction rate was much faster than the diffusion rate away from the interface and favored therefore a multiple stitching architecture, as shown schematically in Fig. 50. Such an interfacial molecular structure does not favor good entanglements with the homopolymer and is mechanically weak.

When the layers of SMA were thinner, the measured fracture toughness increased dramatically. Lee and Char argued that the diffusion rate away from the interface becomes more and more important as the initial layer becomes thinner

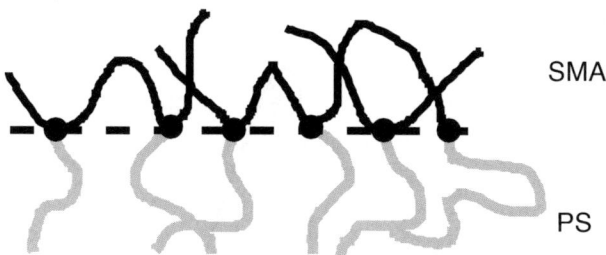

Fig. 50. Possible mechanism by which chains with multiple reactive sites can graft to an interface. This example, which would be typical of a maleic anhydride functionalized polymer reacting on a polyamide, shows on one side end-grafted chains and on the other side of the interface, a loop structure. The effect of this loop structure on the mechanical strength of the interface is not fully clear but loops that are too short will weaken the interface

and favors single stitching where only one reactive group reacts at the interface. These conclusions are supported by experiments at different annealing temperatures which show that, at initial thicknesses of SMA smaller than 30 nm, G_c decreases with the annealing temperature while for thicker initial films, G_c increases with the annealing temperature.

It should be noted, however, that in a later publication [94], Lee and Char pointed out that SMA is significantly more brittle than PS and this could also explain the lower fracture toughness of interfaces where a thick layer of nearly pure SMA is present close to the interface.

A similar study was undertaken by Beck Tan et al. on the adhesion between poly(styrene-r-sulfonated styrene) and poly(2-vinylpyridine). In this case, however, the variable was the mole fraction of sulfonated styrene in the random copolymer [95]. The results of the maximum fracture toughness G_c vs. mole fraction of functional groups are plotted in Fig. 51. The reinforcement shows a very sharp maximum with degree of functionalization consistent with the multiple stitching giving rise to short loops poorly entangled with the homopolymer; however, in this case as well, the bulk properties of PS are modified by the presence of the styrenesulfonic acid and this could contribute to the decrease in G_c at high levels of functionalization.

6.2
Interfaces Between Semicrystalline Polymers

While due to their well-known plastic deformation properties glassy polymers provide excellent model systems for fracture studies, most engineering plastics are semicrystalline. Nevertheless, the molecular mechanisms of reinforcement of interfaces between semicrystalline polymers are much less well understood and the first systematic studies on the subject have only appeared recently [16, 30, 96–99]. The reasons for this are mainly twofold:

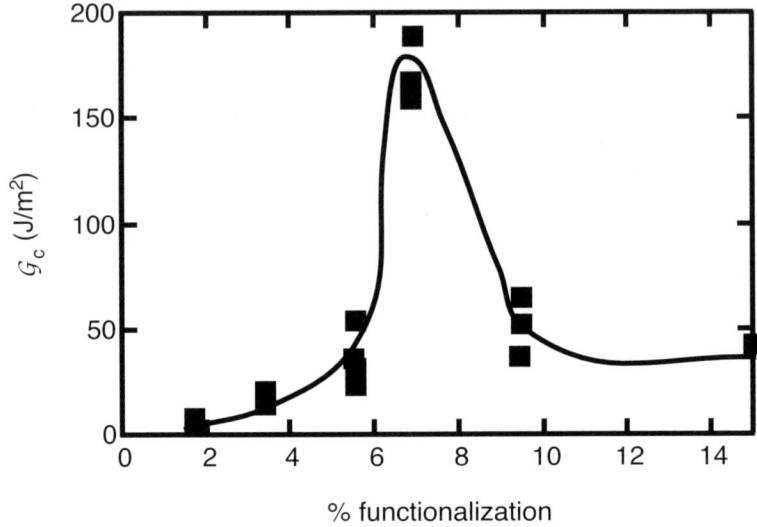

Fig. 51. Fracture toughness of interfaces between poly(2-vinylpyridine) and poly(styrene) containing a small amount of sulfonated PS. G_c is plotted as a function of the mole fraction of PS which is functionalized. Data from [95]

- the polymers that do crystallize have typically a very stereoregular structure and are consequently polymerized by coordination polymerizaton techniques. These synthetic methods typically do not give a very good control of molecular weight and block copolymers are much harder or impossible to synthesize.
- the deformation mechanisms of semicrystalline polymers are much more complicated than those for glassy polymers and typically depend strongly on processing conditions.

Despite those difficulties, some reactive systems have been used to study the reinforcement of interfaces between semicrystalline polymers. Within the context of interfacial fracture, the possibility of crystallization introduces two new important aspects that need to be taken into account when interpreting experimental data:

- the coupling between the formed block and the homopolymer can occur either through entanglement or through incorporation of both chains in the same crystallite.
- the microstructure of the polymer near the interface (highly dependent on thermal treatment) will strongly influence the plastic deformation properties of the polymer near the interface and therefore the fracture toughness.

A large proportion of the industrial applications of reactive compatibilization are between polyolefins and polyamides and several studies have investigated the effect of reactive grafting on the mechanical properties of phase-separated blends of that type; specifically, the improvement in the overall toughness of the blend due to the grafting and the stabilization of the size of the minority phase are well documented. On the specific topic of mechanical reinforcement of planar interfaces, Bidaux et al. [97] and Boucher et al. [16, 96] have both investigated the reinforcement of interfaces between PA-6 (T_m = 225 °C) and PP (T_m = 160 °C) with PP-g-MA chains. In this system a small fraction (1–5%) of functionalized PP chains are dissolved in pure PP and react at the interface with amine-terminated PA-6 chains. While the first study focused on realistic processing conditions (annealing temperatures of a few minutes) and correlated annealing temperature with G_c, the second study focused on correlating the areal density of copolymer chains formed at the interface with fracture toughness. An additional study was recently published on the adhesion between an amorphous polyamide and PP containing a small percentage of MA-functionalized chains [100]. In this case the variables were annealing time, temperature and concentration of functionalized chains in the PP.

In all three studies a clear increase in G_c was observed when the assembly was annealed at least above the melting point of the PP, implying that the functionalized chains indeed need to diffuse to the interface to react with the amine groups of the PA. Boucher et al. [16] and Cho et al. [101] unambiguously proved that a copolymer was formed at the interface as a result of the annealing procedure, by dissolving the PA part of the assembly in a suitable solvent and performing a surface analysis of the PP side. Contact angle measurements showed the presence of PA on the PP surface even for short annealing times [101], while XPS experiments allowed Boucher et al. [16] to determine in a quantitative way the areal density of block copolymer chains present at the interface.

In this series of experiments, Boucher et al. showed that for end-functionalized PP chains (M_n = 22,000) grafted on the amine end-groups of PA-6, $G_c \propto \Sigma^2$ is in excellent agreement with the model presented in Sect. 3. It is interesting to note that, although the presence of crystallites must complicate the deformation process, if the processing conditions are kept identical and the only variable is the areal density of chains, the fracture toughness is accurately predicted by Brown's model. This result was further confirmed by a later study from the same group [102] where reinforcement of the interface between a PP-based blend (containing 70% weight of crosslinked ethylene-propylene rubber particles and 5% of the same PP-g-MA chains used in the previous study) and PA-6 gave an identical scaling of G_c with Σ^2.

The results of both studies are summarized in Fig. 26 and it is clear from the data that the PP-blend system has a higher fracture toughness than the pure PP system for all values of Σ. If one now plots on Fig. 26b the value of σ_{fibril}/σ_c for both systems [using Eq. (24) and the value of $\delta^\ddagger$ obtained from fitting the PP/PA-6 interface data to Eq. (19)], both data sets fall on the same line as one would expect for two systems which have the same interfacial structure but different values of σ_{craze}. The precise values of the crazing stress in a plastic zone are not directly known for PP and for the blend but the yield stress in tension was meas-

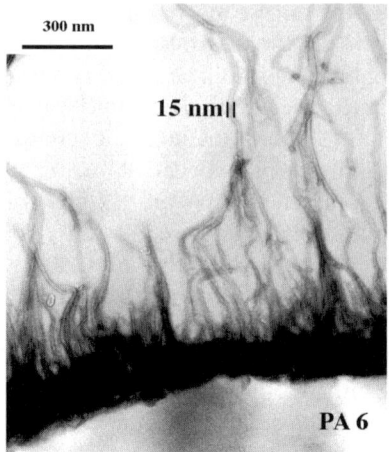

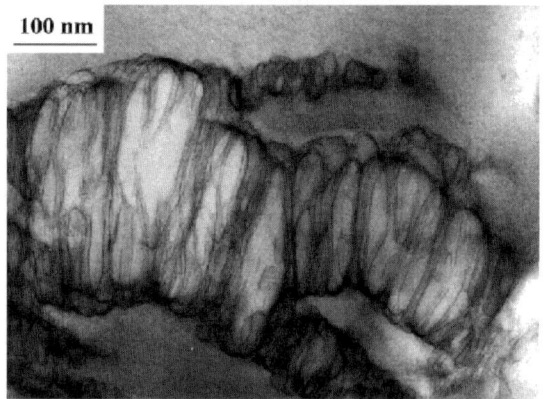

Fig. 52. a TEM micrograph of PP fibrils at the interface between PP and PA-6 [30]. b TEM micrograph of PP fibrils close to the interface between PP-EPDM and PA-6 [102]. These micrographs have been taken very close to the crack tip and show undoubtedly that the crack was preceded by a craze-like structure with a fibril size similar to that observed in glassy polymers

ured at 22 MPa for PP and 4.2 MPa for the PP-blend, in very good agreement with the observed difference in G_c. Furthermore, post-mortem optical and TEM observations of the plastic zone under stress [30, 102, 103] (using a technique of embedding the sample in epoxy under stress prior to staining and microtoming described in Sect. 2) showed that, for both systems, the deformation at the interface occurred by the formation of a localized plastic zone with a fibrillar structure only on the PP side of the interface (see Fig. 52a and 52b), in a very similar way to what has been reported for glassy systems.

An important documented effect specific to polypropylene systems should be mentioned here. Both groups working on the PP-PA-6 reinforcement [96, 97] reported that the interface increased dramatically in toughness when the annealing temperature was above the melting point of PA-6. While the exact nature of this transition is not yet known, interestingly, it cannot be attributed to an increase in areal density of chains since this quantity is independently measured and does not specifically increase with annealing temperature. TEM observations of the plastic deformation zone at the crack tip show the appearance of diffuse deformation ahead of the crack tip in the PP phase in addition to the main cavitational plastic zone [30]. The crystalline microstructure near the interface observed by TEM was however found to be identical for high and low fracture toughness samples implying the existence of a difference in the efficiency of the coupling at the molecular level. Since an opposite effect, i.e. a drop in G_c with increasing annealing temperature, has been reported by Cho et al. in a similar system [101] and attributed to a similar cause, more work is clearly needed to elucidate the reasons of the anomalous behavior.

A different set of experiments provides us with more information on the molecular characteristics needed from the grafted chain. Duchet et al. [98] have investigated the fracture toughness of interfaces between polyethylene (PE) and glass. In order to promote the adhesion between the glass and the PE, the glass surfaces were previously covered with either end-grafted PE or poly(ethylene-co-hexene) chains of various molecular weights or end-grafted long alkane chains (C_4 to C_{30}). For grafted alkane chains, they found that G_c increased with ΣN^2 consistent with a simple chain pullout mechanism and with Eq. (8). For longer PE chains, G_c increased markedly with the molecular weight of the grafted chain, consistent with a fracture mechanism involving the formation of a plastic zone at the crack tip. Two important conclusions could be inferred from their results:
- The most effective reinforcement was obtained with relatively long chains (M_n = 32,500 g/mole) compared to the M_e of PE (~1200 g/mole) which were sparsely grafted (Σ ~ 0.001 chains/nm^2).
- These effective chains only moderately cocrystallized with the HDPE matrix implying that the crystallization of grafted chains together with the matrix chains may not be a necessary condition for a good mechanical reinforcement.

Obviously, these results are in apparent contradiction to results obtained on amorphous systems (good adhesion for M/M_e ~ 5) and counterintuitive (why is cocrystallization not needed), stressing the need for further work on systems with a better controlled molecular structure.

7
Conclusions

From the main results presented in the preceding paragraphs two separate mechanisms of reinforcement of the interface can be distinguished:
1. The formation of individual molecular connectors across the interface. The effect of these connectors can be represented in terms of the two relevant pa-

rameters, i.e. the degree of polymerization of the connector N and the areal density of connectors Σ.

2. The formation of a buffer layer of variable monomer composition that effectively broadens the interfacial region. Although in this case the parameters controlling the mechanical strength of the interface are no longer independent, one can still consider the interfacial width a_I and the degree of polymerization N of the chains in this buffer layer as the controlling parameters.

With this in mind, it is useful to represent the expected fracture mechanisms at the interface with maps. For individual connectors, the fracture mechanisms map can be presented as a function of Σ/Σ^* and N/N_e. This normalization then takes into account two important material parameters of the bulk polymers which will influence the fracture mechanisms map: the crazing stress σ_{craze} (contained in Σ^*) and the entanglement density (contained in N_e).

The map can be divided into four main domains, as shown in Fig. 53. For values of $N/N_e \ll 1$, the dominant mechanism of fracture will always be chain pullout and G_c will be given by Eq. (8). The N/N_e domain, where simple chain pullout is active, decreases however with increasing Σ. The boundary between simple chain pullout and crazing with chain pullout is given by:

$$\frac{\Sigma}{\Sigma^*} = \frac{f_b}{f_{mono} N} \tag{31}$$

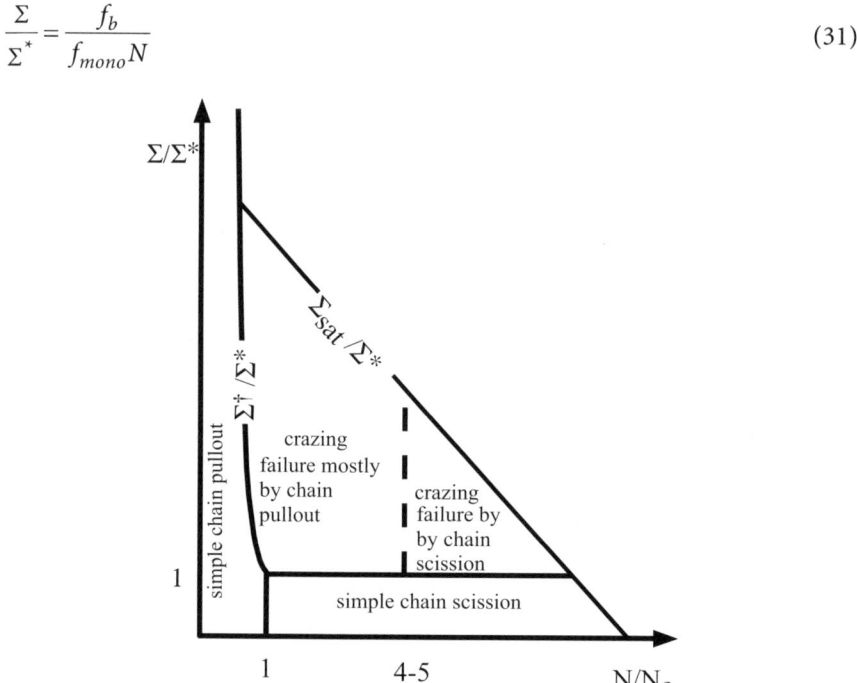

Fig. 53. Fracture mechanisms map for interfaces between glassy polymers reinforced with connecting chains. Failure mechanisms are represented as a function of normalized degree of polymerization N/N_e and normalized areal density of connectors Σ/Σ^*

Similarly, for values of $\Sigma/\Sigma^* < 1$ and $N/N_e > 1$, the only possible failure mechanism is simple chain scission without onset of crazing. These two failure mechanisms give low values of G_c of the order of a few J/m^2 and thus are characteristically observed for weak interfaces.

In order to increase G_c, two strategies are possible: either increasing N/N_e or increasing Σ/Σ^* in order to obtain a failure of the interface by crazing. The crazing failure domain is broadly limited by the triangle shown on Fig. 53, with the left boundary being given by Eq. (31), the lower boundary by the condition $\Sigma/\Sigma^* = 1$ and the right boundary by the value of Σ_{sat}/Σ^* as a function of N/N_e.

The exact form of the right boundary limit, i.e. the limitation due to the maximum number of connecting chains that can be accommodated by the interface, has been discussed in Sect. 4 and will depend on sample preparation conditions. It is clearly, however, always a decreasing function of N which will intersect the two other boundary lines.

At very high values of N, Σ_{sat} will be too low to activate crazing. This condition will be expressed by:

$$\Sigma_{sat} f_b < \sigma_{craze} \tag{32}$$

At very low values of N, the pullout stress at $\Sigma = \Sigma_{sat}$ will be too low to activate crazing as well. This condition will be expressed by:

$$\Sigma_{sat} f_{mono} N < \sigma_{craze} \tag{33}$$

Within the crazing regime one can further distinguish two subregimes: for low values of N/N_e, the dominant failure mechanism within the craze will be chain pullout with some chain scission. The amount of chain scission will increase as N/N_e increases until for a value of N/N_e larger than about 5, the only craze failure mechanism will be chain scission. This regime will give interfaces with the highest values of G_c and the fracture toughness will be given by Eq. (18).

For practical reasons it would be clearly useful to extend the domain of maximum toughness to the largest possible range of values of Σ and N. Since the lower and left boundaries of the domain are system-independent, the only way to extend the crazing/scission domain is to displace the right-side boundary given by Σ_{sat}/Σ^*. Let us examine how this boundary will shift when the two main bulk parameters, σ_{craze} and N_e, are modified. For homologous systems one can expect Σ_{sat} at constant N/N_e to vary as $1/N_e$. Therefore, if the crazing stress of the homopolymers does not change, a more tightly entangled system should give a higher maximum fracture toughness than a weakly entangled system. This is apparent from Fig. 32 where the PMMA system gives higher maximum values of G_c than the PS-based system. However, the crazing stress of the polymer may also change with N_e and should, broadly speaking, increase with decreasing N_e. Some of the advantage of reducing N_e may therefore be offset by an increase in crazing stress.

A similar fracture mechanisms map can also be drawn for interfaces effectively broadened by a buffer layer. In this case we assume that the buffer layer is laterally homogeneous. As shown in Fig. 54, several different fracture mecha-

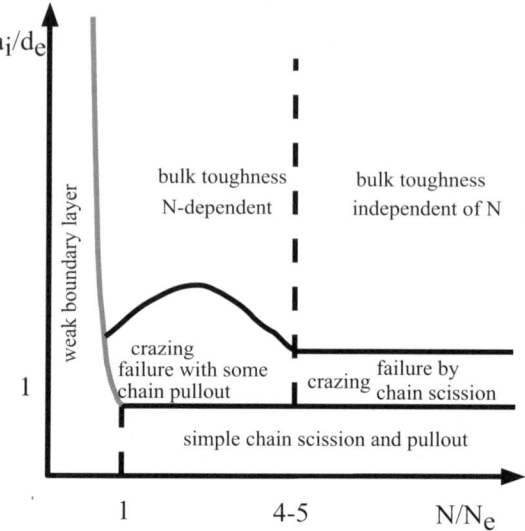

Fig. 54. Fracture mechanisms map for interfaces between glassy polymers as a function of normalized degree of polymerization N/N_e and normalized interfacial width a_I/d_e

nisms can be observed as well. In this case, however, the y-axis is no longer an areal density of connectors but an interfacial width a_I normalized by the average distance between entanglement points d_e.

For low values of N/N_e the interface gives very low values of G_c since the presence of a low molecular weight layer at the interface, regardless of its thickness, will promote disentanglement and prevent a stable craze from forming. This situation is often referred to as a weak boundary layer and is analogous to the results discussed in Sect. 3.6.

For a very narrow interfacial width regardless of the value of N/N_e, the interface will be weak as well. This is the case for strongly immiscible polymers where interpenetration at the interface is insufficient to activate crazing and corresponds to regime I of Fig. 40.

For higher values of a_I/d_e, the failure mechanism becomes crazing. Within the crazing regime one can then define four subregimes:

1. low N/N_e and low a_I/d_e: In this case the failure of the craze is essentially by disentanglement. This is the regime where interfacial width and molecular weight are most closely coupled and where quantitative prediction or modeling is most difficult.
2. low N/N_e and high a_I/d_e: In this regime the toughness of the interface is indistinguishable from that of the bulk polymers, which is, however, N-dependent [104].
3. high N/N_e and low a_I/d_e: This regime corresponds to regime II of Fig. 40 and the fracture toughness G_c is only dependent on interfacial width.

4. high N/N_e and high a_I/d_e: This is the case where a molecular weight independent bulk toughness is recovered.

The precise location of the boundaries between these regimes is not yet completely clear experimentally but it is very likely that for high N/N_e the boundaries should be independent of N and related to the average distance between entanglements, while for low N, the boundaries should be N-dependent and more related to the radius of gyration of the chain, as extensively discussed by Wool [104].

We have shown that the fracture toughness of interfaces between polymers is dependent on the molecular structure at the interface as well as on the bulk properties of the polymers on either side of the interface. This relationship is now relatively well established for glassy polymers and the main results are summarized in Figs. 53 and 54, as well as in Sects 3.2–3.5. However, these results should be used with caution when the polymers on either side of the interface are rubbery or semicrystalline. The stress-transfer mechanisms, and in particular the role of the entanglements, will be very different from those observed for the glassy polymers and only preliminary data are currently available on those systems. In principle, fracture mechanisms maps analogous to those depicted in Figs. 53 and 54 could be drawn for these systems but the relevant parameters are not yet as clearly identified.

Among future challenges in this area of research one can mention:
- The molecular requirements for effective stress-transfer at interfaces with semicrystalline systems.
- The understanding of the interfacial control of crack tip plasticity for systems where the dissipation is no longer confined to a localized cavitational zone but is diffuse or away from the interface.
- The role of the phase angle ψ and of the finite size of the sample on the initiation of plasticity mechanisms. While it is widely recognized that there is such on effect, it is still very poorly understood and will require further close collaboration between the solid mechanics community and the physics and materials science community.

Acknowledgements. EJK wishes to acknowledge support for this work from the MRSEC program of the National Sciences Foundation under Award No. DMR 00-80034. CYH would like to acknowledge the Cornell Center of Materials (CCMR), a Material Research Science and Engineering Center of the National Sciences Foundation (DMR-00-79992) for their support.

References

1. Helfand E, Tagami Y (1972) J Chem Phys 56:3592–3601
2. Hawker CJ (1994) J Am Chem Soc 116:11185–11186
3. Kotani Y, Kamigaito M, Sawamoto M (1998) Macromolecules 31:5582–5587
4. Hawker CJ, Elce E, Dao J, Volksen W, Russell TP, Barclay GG (1996) Macromolecules 29:2686–2688
5. Haddleton DM, Crossman MC, Hunt KH, Topping C, Waterson C, Suddaby KG (1997) Macromolecules 30:3992–3998

6. Odian G (1981) Principles of polymerization. Wiley, New York
7. Kramer EJ (1991) Physica B 173:189–198
8. Mills PJ, Green PF, Palmstrom CJ, Mayer JW, Kramer EJ (1984) Appl Phys Lett 45:957–959
9. Chaturvedi UK, Steiner U, Zak O, Krausch G, Schatz G, Klein J (1990) Appl Phys Lett 56:1228–1230
10. Chaturvedi U, Steiner U, Zak O, Krausch G, Klein J (1989) Phys Rev Lett 63:616–619
11. Payne RS, Clough AS, Murphy P, Mills PJ (1989) Nucl Instrum Methods B 42:130
12. Sokolov J, Rafailovich MH, Jones RAL, Kramer EJ (1989) Appl Phys Lett 54:590–592
13. Genzer JB, Rothman JB, Composto RJ (1994) Nucl Instrum Methods B 86:345
14. Brown HR, Deline VR, Green PF (1989) Nature 341:221–222
15. Kramer EJ (1996) MRS Bulletin 21:37
16. Boucher E, Folkers JP, Hervet H, Léger L, Creton C (1996) Macromolecules 29:774–782
17. Brown HR (1989) Macromolecules 22:2859–2860
18. Hutchinson JW, Suo Z (1991) Adv Appl Mech 29:63–191
19. Erdogan F, Sih CF (1963) J Basic Eng Trans ASME 85:519
20. Rice JR (1988) J Appl Mech 55:98–103
21. Xiao F, Hui CY, Kramer EJ (1993) J Mater Sci 28:5620–5629
22. Creton C, Kramer EJ, Hui CY, Brown HR (1992) Macromolecules 25:3075–3088
23. Kanninen MF (1973) Int J Fracture 9:83–92
24. Brown HR (1990) J Mater Sci 25:2791–2794
25. Xiao F, Hui C-Y, Washiyama J, Kramer EJ (1994) Macromolecules 27:4382–4390
26. Bernard B, Brown HR, Hawker CJ, Kellock AJ, Russell TP (1999) Macromolecules 32:6254–6260
27. Sikka M, Pellegrini NN, Schmitt EA, Winey KI (1997) Macromolecules 30:445–455
28. Washiyama J, Creton C, Kramer EJ (1992) Macromolecules 25:4751–4758
29. Lauterwasser BD, Kramer EJ (1979) Philos Mag A 39:469–495
30. Plummer CJG, Kausch HH, Creton C, Kalb F, Léger L (1998) Macromolecules 31:6164–6176
31. Washiyama J, Kramer EJ, Creton C, Hui CY (1994) Macromolecules 27:2019–2024
32. Washiyama J, Kramer EJ, Hui CY (1993) Macromolecules 26:2928–2934
33. Washiyama J, Creton C, Kramer EJ, Xiao F, Hui CY (1993) Macromolecules 26:6011–6020
34. Brown HR, Char K, Deline VR, Green PF (1993) Macromolecules 26:4155–4163
35. Cho K, Brown HR, Miller DC (1990) J Polym Sci Part B Polym Phys 28:1699–1718
36. Char K, Brown HR, Deline VR (1993) Macromolecules 26:4164–4171
37. Creton C, Brown HR, Deline VR (1994) Macromolecules 27:1774–1780
38. Norton LJ, Smigolova V, Pralle MU, Hubenko A, Dai KH, Kramer EJ, Hahn S, Berglund C, DeKoven B (1995) Macromolecules 28:1999–2008
39. Kramer EJ (1995) Isr J Chem 35:49–54
40. Sha Y, Hui CY, Kramer EJ, Hahn SF, Berglund CA (1996) Macromolecules 29:4728–4736
41. Suo Z, Hutchinson JW (1990) Int J Fracture 43:1–18
42. Xu DB, Hui CY, Kramer EJ, Creton C (1991) Mech Mater 11:257–268
43. Raphaël E, de Gennes PG (1992) J Phys Chem 96:4002–4007
44. Ji H, de Gennes PG (1993) Macromolecules 26:520–525
45. Dai KH, Washiyama J, Kramer EJ (1994) Macromolecules 27:4544–4553
46. Dai CA, Jandt KD, Iyengar D, Slack NL, Dai KH, Davidson WB, Kramer EJ (1997) Macromolecules 30:549–560
47. Brown HR (1991) Macromolecules 24:2752–2756
48. Miller P, Buckley DJ, Kramer EJ (1991) J Mater Sci 26:4445–4454
49. Behan P, Bevis M, Hull D (1975) Proc R Soc London Ser A Mathematical and Physical Sciences A243:525
50. Hui CY, Ruina A, Creton C, Kramer EJ (1992) Macromolecules 25:3948–3955
51. Sha Y, Hui CY, Ruina A, Kramer EJ (1997) Acta Mater 45:3555–3563

52. Donald AM, Kramer EJ, Bubeck RA (1982) J Polym Sci Polym Phys Ed 20:1129–1141
53. Donald AM, Kramer EJ (1982) J Polym Sci Polym Phys Ed 20:899–909
54. Sha Y, Hui CY, Ruina A, Kramer EJ (1995) Macromolecules 28:2450–2459
55. Goodier JN, Field FA (1963) In: Drucker DC, Gilman JJ (eds) International conference on fracture of solids (Metallurgical Society conferences), vol 20. Interscience, New York, pp 103–118
56. Dugdale DS (1960) J Mech Phys Solids 8:100–104
57. Kramer EJ (1997) Plast, Rubber and Compos Process Appl 26:241–249
58. Wool RP, Yuan BL, McGarel OJ (1989) Polym Eng Sci 29:1340–1367
59. Dai CA (1995) PhD thesis, Cornell University
60. Kramer EJ, Norton LJ, Dai CA, Sha Y, Hui CY (1994) Faraday Discuss Chem Soc 98:31–46
61. Gent AN, Schultz J (1972) J Adhes 3:281–294
62. Maugis D, Barquins M (1978) J Phys D Appl Phys 11:1989–2023
63. Passade N, Creton C, Gallot Y (2000) Polymer 41:9249–9263
64. Brown HR, Krappe U, Stadler R (1996) Macromolecules 29:6582–6588
65. Lake GJ, Lindley PB (1965) J Appl Polym Sci 9:1233–1251
66. Dai CA, Kramer EJ, Washiyama J, Hui CY (1996) Macromolecules 29:7536–7543
67. Schnell R, Stamm M, Creton C (1998) Macromolecules 31:2284–2292
68. Schnell R, Stamm M, Creton C (1999) Macromolecules 32:3420–3425
69. Brown HR (1999) in Macromolecules 34:3720-3724
70. Benkoski J, Kramer EJ, Fredrickson GH (2000) J. Polym. Sci. Polym. Phys. in press Ed.
71. Bernard B, Brown HR, Russell TP, Hawker CJ (1996) Polym Mater Sci Eng 29:155–156
72. Kulasekere R, Kaiser H, Ankner JF, Russell TP, Brown HR, Hawker CJ, Mayes AM (1996) Macromolecules 29:5493–5496
73. Kulasekere R, Kaiser H, Ankner JF, Russell TP, Brown HR, Hawker CJ, Mayes AM (1996) Physica B 221:306–308
74. Smith GD, Russell TP, Kulasekere R, Ankner JF, Kaiser H (1996) Macromolecules 29:4120–4124
75. Dai CA, Osuji CO, Jandt KD, Dair BJ, Ober CK, Kramer EJ (1997) Macromolecules 30:6727–6736
76. Edgecombe BD, Stein JA, Fréchet JMJ, Xu Z, Kramer EJ (1998) Macromolecules 31:1292–1304
77. Edgecombe BD, Fréchet JM, Xu Z, Kramer EJ (1998) Chem Mater 10:994–1002
78. Xu Z, Kramer EJ, Edgecombe BD, Fréchet JMJ (1997) Macromolecules 30:79–84
79. Dai C-A, Dair BJ, Dai KH, Ober CK, Kramer EJ, Hui CY, Jelinski LW (1994) Phys Rev Lett 73:2472–2475
80. Noolandi J, Shi AC (1995) Phys Rev Lett 74:2836
81. Milner ST, Fredrickson GH (1995) Macromolecules 28:7953–7956
82. Shull KR, Kramer EJ, Hadziioannou G, Tang W (1990) Macromolecules 23:4780–4787
83. Paul DR, Barlow JW (1984) Polymer 25:487–494
84. ten Brinke G, Karasz FE, MacKnight WJ (1983) Macromolecules 16:1827–1832
85. O'Shaughnessy B, Sawhney U (1996) Macromolecules 29:7230–7239
86. Fredrickson GH, Milner ST (1996) Macromolecules 29:7386–7390
87. O'Shaughnessy B, Vavylonis D (1999) Macromolecules 32:1785–1796
88. Fredrickson GH (1997) J Chem Phys 106:2458–2468
89. Jiao J, Kramer EJ, de Vos S, Koning C (1999) Polym Commun 40:3585–3588
90. Jiao J, Kramer EJ, de Vos S, Möller M, Koning C (1999) Macromolecules 32:6261–6269
91. Lyu S-P, Cernohous JJ, Bates FS, Macosko CW (1999) Macromolecules 32:106–110
92. Boucher E (1995) Thèse de doctorat, Université Paris VI, France
93. Lee Y, Char K (1994) Macromolecules 27:2603–2606
94. Lee Y, Char K (1998) Macromolecules 31:7091–7094
95. Beck Tan NC, Peiffer DG, Briber RM (1996) Macromolecules 29:4969–4975
96. Boucher E, Folkers JP, Creton C, Hervet H, Léger L (1997) Macromolecules 30:2102–2109

97. Bidaux JE, Smith GD, Bernet N, Manson JA, Hilborn J (1996) Polymer 37:1129–1136
98. Duchet J, Chapel JP, Chabert B, Gerard JF (1998) Macromolecules 31:8264–8272
99. Xue YQ, Tervoort TA, Lemstra PJ (1998) Macromolecules 31:3075–3080
100. Cho K, Seo KH, Ahn TO, Kim J, Kim KU (1997) Polymer 38:4825–4830
101. Cho K, Li F (1998) Macromolecules 31:7495–7505
102. Kalb F, Léger L, Creton C, Plummer CJG, Marcus P, Magalhaes A (2000) Macromolecules 34:2702-2709
103. Kalb F (1998) Thèse de doctorat, Université Paris VI, France
104. Wool RP (1995) Polymer interfaces. Hanser Verlag, Munich
105. Dai KH, Kramer EJ (1994) Polymer 35:157–161
106. Dai KH, Norton, LJ, Kramer, EJ (1994) Macromolecules 27:1949–1956

Editor: Prof. H.-H. Kausch
Received: Novembe 2000

Recent Progress in Gel Theory: Ring, Excluded Volume, and Dimension

Kazumi Suematsu

Institute of Mathematical Science, Ohkadai 2-31-9, Yokkaichi, Mie 512-1216, Japan
e-mail: suematsu@mint.ocn.ne.jp

Recent developments in gel research are reviewed with emphasis on the gel point problem. We will describe in due course how the gel point equation can be deduced from first principles. First we review briefly the industrial development of gel science in Japan (Sect. 1) and a central aspect of the classical theory of gelation (Sect. 2). In Sect. 3, we survey the progress on the excluded volume problem from the author's point of view. In all respects, this theme is, now, too biased to physics and hence beyond the scope of this review; while it is an essential subject to understand the nature of the gel point. Regarding the excluded volume problem, a recent interesting idea is the screening effect. This notion of screening is a different interpretation of the Flory excluded volume theory, but takes us into advanced physics. For instance, the behavior of a branched molecule in the melt becomes comprehensible in a natural fashion. In Sect. 4, we mention cyclization in branching media. Like the problem of volume exclusion, the cyclization problem has not been solved rigorously. The most troublesome aspect with cyclization is that there is no way to enumerate the combinatorial number of branched molecules with rings. On the other hand, the mathematical framework for the general solution has already been given. In this article we will mention the limiting solutions of $C \rightarrow \infty$ for real systems and of $d \rightarrow \infty$ for the lattice model. What is important is that these limiting solutions are by no means useless, fictitious entities, but have real meanings. By analogy with the $f = 2$ case, we can put forward the general relation, $[\Gamma] \cong constant$ for gelation conditions, where $[\Gamma]$ represents the total ring concentration; this is the basic premise of the gel point theory developed in Sect. 6. Through these analyses, essential differences between real gelations and the percolation model are brought into sharp relief (Sect. 5). The gel point theory starts from the obvious equality: $D_c = D(inter) + D(ring)$, where D_c represents the gel point, $D(inter)$ the extent of the intermolecular reaction alone at the gel point and $D(ring)$ the corresponding quantity of cyclization. Then, according to some definitions, fundamental equalities for gel points can be deduced for all the models of real systems and the percolation model. The problem of seeking a gel point for a given system thus reduces to the problem of finding a solution for the corresponding fundamental equality. To solve the equalities, we introduce two main assumptions: (1) random distribution of cyclic bonds, and (2) that the ring distribution functions can be expanded about $D_c = D_{co}$, where D_{co} is the gel point for the ideal tree model. Under these assumptions, we can derive analytical expressions for gel points as functions of γ ($= 1/C$: the reciprocal of an initial monomer concentration), κ (mole ratio of B-type functional units to A-type functional units), and d (space dimension); that is, $D_c = G(\gamma, \kappa, d)$. In Sect. 7, the theoretical equations thus obtained are compared with experiments. The result shows that the theory recovers well the points observed by Flory, Weil, and Gordon in all the regimes of $\kappa = 1 \sim 2$. The corresponding expression for the percolation model is found to agree well with simulation experiments in high dimensions, but fails in low dimensions. The discrepancy in low dimensions is analyzed in light of the critical dimension concept. One possible explanation is that the above-mentioned assumptions (1) and (2) do not work below $d_c = 8$.

Keywords: Branching process, Gelation, Cyclization, Excluded volume effects, Dimension, Concentration invariant, High concentration expansion, High dimension expansion, Gel point, Percolation threshold, Critical dimension.

1	**Introduction**	141
2	**Classical Theory of Gelation**	143
3	**Excluded Volume Problem**	146
3.1	Excluded Volume Effects of Chain Molecules	147
3.2	Excluded Volume Effects of Branched Molecules	150
3.3	Numerical Estimation	152
3.4	Conclusions	154
4	**Ring Formation**	154
4.1	Cyclization in Real Systems	155
4.1.1	R-A_f Model	155
4.1.2	R-A_g + R-B_{f-g} Model	158
4.1.3	A_g-R-B_{f-g} Model	158
4.1.4	Physical Aspects of Asymptotic Equations	163
4.1.5	Another Solution	165
4.2	Cyclization on Lattice	168
4.2.1	Numerical Work	168
4.2.2	Ring Distribution Functions on Lattices	169
4.3	Conclusions	173
5	**Real Gelation and Percolation Process**	174
5.1	Conclusions	178
6	**Estimation of Gel Point**	178
6.1	Experimental Determination of Gel Point	178
6.2	Numerical Calculation	180
6.3	Analytical Approach	183
6.3.1	Gel Point in Real Polymer Solutions	183
6.3.2	Critical Dilution	189
6.3.3	Threshold in the Percolation Model	190
6.4	Conclusions	193
7	**Comparison with Experiments**	194
7.1	Real Systems	194
7.1.1	*gem*-Substituent Effects	197
7.1.2	Without *gem*-Substituent Effects	199

7.1.3	Comparison with Viscoelastic Method 201
7.2	Percolation Model 203
7.2.1	Pure Bond Percolation ($p_s = 1$) 205
7.2.2	Bond Percolation ($p_s \neq 1$) 205
7.2.3	Critical Dimension 206
7.3	Conclusions 208
8	**Expectations of Future Theories** 209
	References 211

Abbreviations

a	attractive interaction between molecules; sum of radii of two interacting molecules
b	bond number; molar excluded volume
A	Helmholtz free energy; probability generating function
A_2	second virial coefficient
$\underset{\sim}{A}$	square matrix
$\mathcal{B}$	function of D
c	weight concentration [gl^{-1}]
C, C^*	initial monomer concentration, and boundary concentration
C'_M	coefficient regarding the fifth power law of volume exclusion
$\mathcal{C}$	volume of a d-dimensional sphere with a radius of unity
d, d^*, d_c	dimension, boundary dimension, and critical dimension
D	extent of reaction of functional units
D_c, D_{co}	gel point, and the classical gel point for the tree model
$f, g, f\text{-}g$	number of functional units
FU	functional unit
G	gel point as a function of γ, κ and d
G', G''	dynamic storage and loss modulus, respectively
I, I'	rate constants of real systems and of the percolation process, respectively
ℓ	bond length
$\mathcal{L}$	one dimensional length of a lattice
m	number of unreacted functional units
n	mole number, degree of polymerization of polymer molecules as solvent
M	probability generating function; moment generating function; molecular mass
M_0	total numbers of A-type monomer units
N	degree of polymerization of polymer molecules; number of functional units
N_b	number of animals with b bonds
N_0	total numbers of B-type monomer units

N_{R_j}	number of j-rings; $[N_{R_j}] = N_{R_j}/V$
$N(A)_j$	number of A functional units in the jth generation
$N(B)_j$	number of B functional units in the jth generation
$N_x(U)_j$	number of branching units in the jth generation on an x-cluster
$\mathbf{N}_j$	2×1 matrix for the numbers of A or B functional units in the jth generation
$\mathcal{N}$	number of monomers on one dimensional length
p	bond probability
p	bond probability for reversible gelation
p_R	probability of a functional unit being occupied by cyclic bonds
p_s	probability of a given site being a monomer unit
$p\{ring\ j\}$	transition probability of j-chains to j-rings per unit bond formation
$p\{ring\ j\}$	transition probability of intermolecular reaction
P	pressure
P_{cy}	probability of a j-chain to close a ring
$P(p)$	probability of a given monomer unit belonging to an infinite molecule
$\mathcal{P}$	probability of one end of a j-chain entering a small volume v
r	atomic radius, end-to-end distance
$\langle r_N^2 \rangle$	mean end-to-end distance of N-chain, mean gyration radius of N-branched molecule
S_d	surface area of a d-dimensional sphere
$S(D)$	mean cluster size at D
u	rate of backward reaction
$U(r)$	potential energy at a distance r
v	volume fraction; excluded volume parameter; small volume around a functional unit
v_0	real volume calculated from an atomic radius
v_{coil}	volume occupied by a polymer molecule
v_L	rate of intermolecular reaction
v_{R_j}	rate of cyclization of j-chains; $v_R = \sum_{j=1}^{\infty} v_{R_j}$
V	system volume
V_p	*molar volume of polymers*
V_s	*molar volume of solvents*
w	weight fraction of rings
X	ring concentration as a function of D
Z	partition function; ring concentration as a function of D
α	Flory expansion factor
χ	interaction parameter between molecules
ε	small number
ϕ_j	number of chances of cyclization for j-chains
γ	exponents of the number of self-avoiding walks; reciprocal of an initial monomer concentration
γ_c	critical dilution value

Γ	gamma function
$[\Gamma]$	total ring concentration in *mol/l*, $[\Gamma] = \sum_{j=1}^{\infty}\left[N_{R_j}\right]$
φ_j	relative cyclization frequency of *j*-chains to intermolecular reaction
κ	mole ratio of B-type functional units to A-type functional units
λ	eigenvalue
ν	exponent of the radius of gyration
ξ	number of bonds per repeating unit
Π	osmotic pressure
θ	exponent of the number of lattice animals
Θ	Flory temperature where attractive force and repulsive force exactly balance
ϱ	exponent expressing solvent size
σ	reciprocal of the classical critical point
ψ	generating function of number of branching units
ζ	dummy variable

1
Introduction

The science of gel materials is a rapidly developing field. This new branch of science dates back to the 1930s when Carothers' series of papers [1] on synthetic polymers was published; 10 years later, in 1941, the first theoretical work by Flory appeared [2]. Setting an independent Chapter on *Polyfunctional Reactions and Non-linear Polymers* in his review article entitled '*Polymerization*' published in 1931, Carothers remarks, "polymerization is not limited to bifunctional compounds; if one of two reactants contains polyfunctional groups, the product will be not a simple chain, but a more complicated structure; the reaction first leads to the formation of a soft, soluble, thermoplastic resin; this, on being heated, further yields a hard, insoluble resin which is completely lacking in thermoplasticity." Today we know that the observed loss of thermoplasticity is intimately connected with the appearance of an infinite network called a gel molecule. Thanks to the remarkable insight of Carothers, today we can still enjoy continuous excitement with this growing science.

The research and development in this field in Japan began in 1930s before World War II: The development in Japan progressed in a rather different way, where like the situation in Europe of that time, owing to the extreme shortage of materials through before and during World War II, the nation's policy inevitably concentrated on the production of the commodities such as clothes. As for clothes, for most Japanese all fabrics had to be like silk; smooth to the touch, vivid to dye, lustrous and colorful. Along this requirement the textile called Vinylone with silky touch and high moisture absorption was invented [3] and soon manufactured on an industrial scale in 1939 by means of the addition polymerization of vinyl acetate followed by hydrolysis of the acetyl moiety, then treated with formalin solution. Although the polymerizability of vinyl acetate itself was

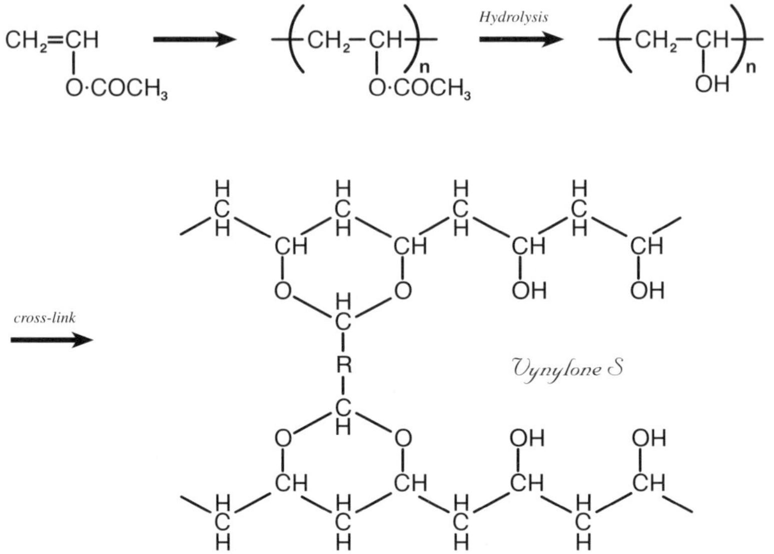

Fig. 1. Synthetic scheme of Vinylone S. The first synthetic fiber in Japan

discovered by the German chemists Herrmann and Haehnel [4,5] in 1927, research and development after that was pushed forward in Japan through 1930s under the leadership of a Japanese engineer Lee. This new textile met, if not sufficiently, the people's need. Really, for the Japanese of those days, Vinylone was almost equivalent to clothes. It was straightforward to convert this material into the three-dimensional network resin called "Vinylone S" using a small amount of dialdehyde as cross-linker in place of formalin; thus leading to the first industrial gel material made in Japan; today, Japan is the sole manufacturer of Vinylone in the world.

When we look back again at those works done in the early times, we note that nothing was suggested to make one foresee the present great prosperity of gel science, but that the essential characteristics of gel materials, that afterwards attracted so many talented researchers to this new frontier, were clearly manifested; e.g., sudden change occurring from fluid sol to infusible gel (sol-gel transition); most of the unreacted functional units (FU's) being still present inside a gel (divergence of fluctuation at the gel point); resins being often amorphous in contrast to linear polymers (network structure). These observations of the early researchers became the basis for the theoretical development of modern gel science.

The main purpose of this article is to introduce the recent advances concerning the gel point problem. The author would like to mention not only new approaches, mainly those developed in the author's laboratory, but also those limitations with a great expectation to future theories, since the author believes that all theories are only temporary and will be replaced by more general and comprehensive ones.

2
Classical Theory of Gelation

With deep physicochemical insight, Flory carried gel science further [6]. To translate the complicated structure of gel molecules into comprehensible mathematical languages, he introduced three main assumptions: (i) equireactivity of functional units (FU's), (ii) no ring formation, and (iii) no excluded volume effects, thus simplifying the involved network structure to a plain tree model. Under these assumptions, Flory was able to deduce, using probabilistic arguments, the important two quantities, the gel point expression and the size distribution; in doing so, he laid the foundation of the classical theory of gelation.

The Flory theory was refined with greater generality into a sophisticated mathematical framework by Good [7] who independently established the beautiful theory of the branching process, later to be called the cascade formalism. Let us take a look at the power of this mathematical method. Consider the branching reaction of no rings. The essence of this formalism is to write down the generating function [8–10]:

$$M(\theta) = \theta F_0 \left(\theta F_1 \left(\theta F_2 \left(\theta F_3 (\cdots) \right) \right) \right). \tag{1}$$

Consider the branching process of the $A-R-B_{f-1}$ model. In this model, clearly $F_0 = F_1 = F_2 = \cdots \equiv F$. Thus Eq. (1) can be written as

$$u(\theta) = \theta F(u(\theta)). \tag{2}$$

Moreover, one can put

$$M(\theta) = \sum_{x=1}^{\infty} p_x \theta^x, \tag{3}$$

where p_x is the number fraction of x-clusters. From Eqs. (1)–(3), many important quantities such as the weight average degree of polymerization $(DP)_w$ and the Z average degree of polymerization $(DP)_Z$ can be deduced. For instance, $(DP)_w$ can be calculated via

$$(DP)_w = \left[\frac{\partial}{\partial \theta} \left(\theta \frac{\partial M}{\partial \theta} \right) \middle/ \left(\frac{\partial M}{\partial \theta} \right) \right]_{\theta=1} = \frac{1 - F'^2 + FF''}{(1 - F')^2}, \tag{4}$$

where the symbol ' denotes the differentiation with respect to u. An important point is to note the equivalency of Eq. (3) with the equation:

$$F(u) = (1 - D + Du)^{f-1}, \tag{5}$$

where D is the extent of reaction of B-type FU, which is equivalent to the ratio of the total bond number to the possible bond number. With Eqs. (4) and (5), one immediately recovers the Flory result:

$$(DP)_w = \frac{1-(f-1)D^2}{[1-(f-1)D]^2}. \tag{6}$$

Now let us make a slight modification of Eq. (3) (which is due to Kajiwara [11]):

$$A(\{\theta\}) = \sum_{x=1}^{\infty} p_x \theta_1^x \theta_2^{\psi(\zeta_j)}, \tag{7}$$

where $\psi(\zeta_j) = \sum_{j=1}^{x} N_x(U)_j \zeta_j$, $N_x(U)_j$ denotes the number of branching units in the jth generation on an x-tree, ζ_j is a dummy variable, and

$$A(\{\theta\}) = (1-D+D \cdot A(\{\theta\}))^{f-1} \theta_1 \theta_2^{\zeta_j}. \tag{8}$$

With the recurrent relationship $A_j = \theta_1 \theta_2^{\zeta_j}(1-D+DA_{j+1})^{f-1}$ in mind,

$$[\partial A/\partial \theta_2]_{\theta_2=1} = \sum_{j=1}^{\infty} A \left[\frac{(f-1)DA}{1-D+DA}\right]^{j-1} \zeta_j. \tag{9}$$

Following Good, we apply the Lagrange theorem to Eq. (9) to get

$$[\partial A/\partial \theta_2]_{\{\theta\}=1} =$$

$$\sum_{x=1}^{\infty} \frac{1}{x!} \frac{d^{x-1}}{dA^{x-1}} \left\{(1-D+DA)^{(f-1)x} \sum_{j=1}^{\infty} A \left[\frac{(f-1)DA}{1-D+DA}\right]^{j-1} \left[1+\frac{(1-D)(j-1)}{1-D+DA}\right]\right\}_{A=0} \zeta_j.$$

With the aid of the expression of the number fraction of x-clusters

$$p_x = \frac{(fx-x)!}{x!(fx-2x+1)!} D^{x-1}(1-D)^{fx-2x+1},$$

we have

$$[\partial A/\partial \theta_2]_{\{\theta\}=1} = \sum_{x=1}^{\infty} p_x \sum_{j=1}^{x}(f-1)^{j-1}\{(f-2)j+1\}\frac{\binom{(f-1)x-j}{(f-2)x}}{\binom{(f-1)x}{x}} \zeta_j, \tag{10}$$

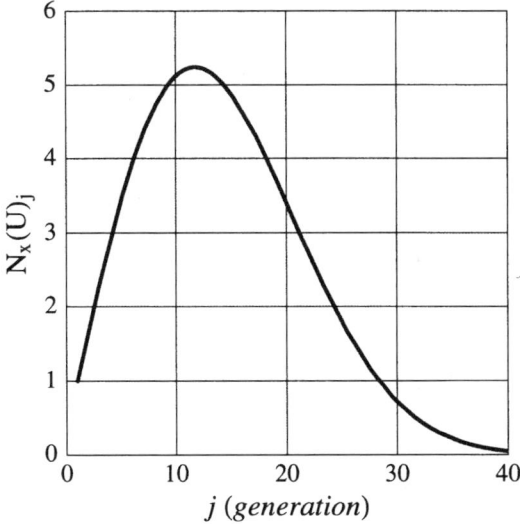

Fig. 2. Molecular profile of an x-cluster formed by polycondensation

while by Eq. (7)

$$[\partial A/\partial \theta_2]_{\{\theta\}=1} = \sum_{x=1}^{\infty} p_x \sum_{j=1}^{x} N_x(U)_j \zeta_j. \tag{11}$$

Comparing termwise the coefficients of ζ_j of Eqs. (10) and (11), we have the expression for the number of branching units in the jth generation on an x-tree [11]:

$$N_x(U)_j = (f-1)^{j-1}\{(f-2)j+1\}\frac{\binom{(f-1)x-j}{(f-2)x}}{\binom{(f-1)x}{x}}, \tag{12}$$

which satisfies $N_x(U)_1 = 1$ for all f's (≥ 2) and x's (≥ 1), as expected.

Eq. (12) represents an average profile for x-clusters (Fig. 2), and was first found by Kurata for the R – A_f model through the combinatorial argument [13]. In Fig. 2 is shown a typical cluster profile for an x = 100 tree as a function of the generation [14]. The derivation of Eq. (12) is one of major accomplishments of the classical gel theory.

In return for introducing the basic assumptions (i) to (iii), the classical theory thus furnished the beautiful mathematical formalisms for (1) the gel point, (2) the molecular weight distribution in sol phase, (3) the gel fraction, and (4) the molecular profile as mentioned above. At first sight, the classical theory may appear to be an unrealistically oversimplified model because of the neglect of ring formation and excluded volume effects. For this reason, it has been long criticized by physicists. As deeper understanding gradually permeates, however, new aspects of the classical theory became manifest themselves. Today, we have come to understand that the classical theory actually dealt with a hypothetical limit of infinite concentration ($C \to \infty$, Sect. 7) where all ring formations and excluded volume effects should vanish rigorously. It has established a standard state of the gel science equal to the relationship of the ideal gas to the real gas. Nowadays, to learn the central concept of the classical theory is an essential step for young researchers to make contributions in this field.

3
Excluded Volume Problem

The notion of excluded volume dates back to the van der Waals formulation for the behavior of real gases [15], the equation of which reads

$$\left(P + a(n/V)^2\right)(V - bn) = nRT, \qquad (13)$$

where a represents the interaction energy between molecules and b the molar excluded volume of a gas molecule. In the classical mechanics, an atomic image is captured as a hard sphere, an incompressible and elastic object. Within the framework of this classical picture, it can be proven that the constant b is exactly equal to four times the real volume v_0 calculated with the atomic radius r,

$$b = 4 N_A v_0.$$

Although atomic radii are roughly proportional to the atomic numbers, in accord with the hard sphere model of atoms, such a classical picture is not consistent with our observations. We know today that atomic radii are actually mean effective radii variable depending on surrounding electric fields and the nature of chemical bonds, which is exemplified in the example of nitrogen, N, atom (atomic number 7; $r \cong 1.55$) and oxygen, O, (atomic number 8; $r = 1.4 \sim 1.5$), namely $r_{Nitrogen} > r_{Oxygen}$, in reverse order to the atomic numbers, an incomprehensible aspect of atomic entity being manifested [16]. When the excluded volume concept of gas molecules is transposed to polymer science, the same problem should arise. Under such circumstances, it is inconceivable that someday one can acquire a complete theory for the excluded volume problem; indeed, modern science does not take such a microscopic standpoint, but instead seeks a solution of more simplified models, namely, the beads model, the lattice model, the hard rods model, and so forth. Even for such drastically simplified models, the excluded volume problem of polymer molecules cannot be solved exactly. Although in this strict sense we are far away from the final end, recent developments in this field are unveiling most of the essential features of the excluded volume problem.

The excluded volume of large molecules changes from solvents to solvents, depending on concentration, temperature and so forth. Now expand Eq. (13) with respect to $\varepsilon = (n/V) = 0$ to yield:

$$P = RT\varepsilon + bRT\left(1 - \frac{\Theta}{T}\right)\varepsilon^2 + 6b^2\varepsilon^3 + \cdots. \tag{14}$$

with $\Theta = a/bR$. The coefficient of the second term expresses two body interactions. At $T = \Theta$ the attractive force and the repulsive force exactly cancel each other out. This is the notion of the Boyle temperature in which the two body interactions apparently vanish. Now convert the variables as $P \to \Pi$, $n/V \to c/M$, and we have the corresponding expression of the osmotic pressure for the macromolecular solution:

$$\Pi = RT\left(\frac{c}{M}\right) + bRT\left(1 - \frac{\Theta}{T}\right)\left(\frac{c}{M}\right)^2 + 6b^2\left(\frac{c}{M}\right)^3 + \cdots. \tag{15}$$

with c being the weight concentration $[g \cdot l^{-1}]$. The first term comprises the van't Hoff equation [17]. If we deal with a dilute polymer solution, the terms higher than the third may be disregarded. Thus, the excluded volume problem focuses in substance on the study of the second term. A major difference of polymer molecules from gas molecules is in their geometrical profiles, the latter being well approximated with a spherical model, whereas the former are composed of a large number of monomer units thus forming a very long string molecule. This difference in geometry is the essence of excluded volume effects in polymer molecules. Thus, it follows that the excluded volume problem in polymer molecules deals with the interaction among monomers within the same molecule.

3.1
Excluded Volume Effects of Chain Molecules

The excluded volume problem of polymer chains was taken up early in 1943 by Flory [6]. His arguments based on the chemical thermodynamics brought the conclusions: (i) the existence of the Flory point (Θ point) where two body interactions apparently vanish, and (ii) that in non-solvent state chains behave ideally.

Afterwards, his central concept was recast by physicists into a more intelligible mathematical language [18–20]. According to their arguments, consider a polymer molecule in the dilute limit, and suppose that the molecule obeys the Gaussian statistics. Then give individual configurations the weighting factor g_{config} depending on the force of the pair interaction. The resulting configurational ensemble will be that of an excluded volume chain. This is the essence of the Flory excluded volume theory. Now all we should do is to write down the end-to-end distance probability distribution in terms of a partition function Z:

$$Z = S_d P(r) dr = Z_{Gauss} Z_{Potential}$$

$$= \left(\frac{d}{2\pi\langle r_N^2 \rangle}\right)^{\frac{d}{2}} S_d \exp\left(-\frac{d}{2\langle r_N^2 \rangle}r^2\right) \sum_{all\ config} g_{config} \cdot \exp\left(-\frac{U(r)}{kT}\right) dr, \quad (16)$$

where $U(r)$ denotes a local potential energy among monomer pairs in specific configuration, and S_d a surface area of a d-dimensional sphere. In the simple hard sphere model (an athermal model) to which $\Theta = 0$ corresponds, $U(r)$ has the form:

$$U(r) = \begin{cases} \infty & (r \leq a) \\ 0 & (r > a) \end{cases}$$

with a being a sum of radii of two interacting molecules. More realistically, $U(r)$ should be taken as a monotonous function of r. The partition function of the potential energy terms is summed over all configurations which have the common end-to-end distance, r. $Z_{Potential}$ measures a fraction of configurations without excluded volume effects, while Z satisfies:

$$\int_0^\infty Z dr = \int_0^\infty S_d P(r) dr = 1.$$

Unfortunately, one cannot evaluate individual potential energies over all configurations. To resolve this difficulty, the following approximation is introduced:

$$\sum_{all\ config} g_{config} \exp\left(-\frac{U(r)}{kT}\right) \Rightarrow \exp\left(-\frac{\bar{U}}{kT}\right), \quad (17)$$

where

$$\bar{U} = \frac{1}{2} kT N (\mathcal{N}/\mathcal{L})^d \cdot v,$$

$\mathcal{N}$ is the number of monomers on a one dimensional length, $\mathcal{L}$ and satisfies $N = \mathcal{N}^d$, and v is a dimensionless coefficient called an excluded volume parameter. Or one can write

$$\bar{U} = \frac{1}{2} kT \left(N^2 / v_{coil}\right) \cdot v,$$

where v_{coil} expressing a polymer volume of the form:

$$v_{coil} = \frac{2\pi^{d/2}}{d\Gamma(\frac{d}{2})} r^d.$$

The approximation Eq. (17) summarizes the essence of the Flory theory, where local fluctuations are averaged out and replaced with a mean potential $\overline{U}$ [19–20].

The partition function is correlated with the Helmholtz free energy via

$$A = -kT \log Z . \tag{18}$$

The force balance between the attractive force (entropy term) and the repulsive one (potential term) is attained at the most probable configuration corresponding to the minimum of the free energy, A. To find the solution, differentiating Eq. (18) with respect to r, and putting $\partial A/\partial r = 0$, we have

$$\frac{N^2}{r^{d+1}} v \cong \frac{r}{\langle r_N^2 \rangle}, \quad \text{for a large } N. \tag{19}$$

Recall that we have made an assumption of the Gaussian statistics for $P(r)dr$, so that $\langle r_N^2 \rangle = N\ell^2$. Substituting this into Eq. (19), we have

$$r \propto N^{\frac{3}{d+2}}. \quad (N \to \infty) \tag{20}$$

By the assumption, $r \propto N^\nu$ for the large N limit, and thus

$$\nu = \frac{3}{d+2}, \tag{21}$$

which was put forward by Fisher on the basis of the Flory argument [18].

The Flory theory, to be sure, ignores the clustering of molecules. Indeed, his theory can be regarded as a polymer version of the van der Waals formulation where no concentration fluctuation is taken into account. In spite of its simplified argument, the Flory theory includes profound physics and agrees remarkably with experiments, giving exact values $\nu = 1$ for $d = 1$ and $\nu = 3/4$ for $d = 2$, respectively, and giving $\nu = 3/5$ for $d = 3$ very close to the observed value 0.588 [21]. The Flory theory was challenged by many theoretical physicists to improve on the mean-field character. To our surprise, more sophisticated mathematical treatments [22–23] as well as numerical calculations [24–25] arrived at the same conclusion shown by Eq. (21). To date, no one has so far surpassed this achievement of Flory. A prominent consequence of the Flory-Fisher theory is that Eq. (21) predicts a critical dimension [20] above which the Gaussian behavior appears. For the present case, the Gaussian exponent $\nu = 1/2$ is recovered at $d = 4$ implying the critical dimension, $d_c = 4$; the classical behavior is expected to work for all dimensions of $d \geq 5$. The Flory-Fisher prediction was confirmed by simulation experiments on lattices [26].

3.2
Excluded Volume Effects of Branched Molecules

A major difference of branched molecules from chain molecules is that more units are bound together and compressed into a very narrow space around the center of gravity. Hence, an immediate supposition is that in order for the monomer-monomer interaction to balance with the monomer-solvent interaction and the entropy force, and for the excluded volume effects to vanish, more attractive force between monomers are needed than is the case of chain molecules. Now we will focus our attention on concentrated systems such as non-solvent systems. An interesting idea is the influence of 'solvent' size on the osmotic pressure (screening effect) [19].

Consider a pair molecules in a polymer melt, and imagine a hypothetical situation of the osmotic pressure of polymer molecules; namely we replace solvent molecules with polymer molecules, and polymer molecules with solvent molecules. Let ν be the mole number and υ the volume fraction, and statistical thermodynamics gives

$$T\Delta S = -RT\left(\nu_s \log \upsilon_s + \nu_p \log \upsilon_p\right), \tag{22}$$

$$\Delta H = RT\chi \nu_s \upsilon_p \tag{23}$$

the subscript s and p denoting solvent and polymer respectively, and

$$\upsilon_s = \frac{\nu_s}{\nu_s + n\nu_p}; \quad \upsilon_p = \frac{n\nu_p}{\nu_s + n\nu_p}.$$

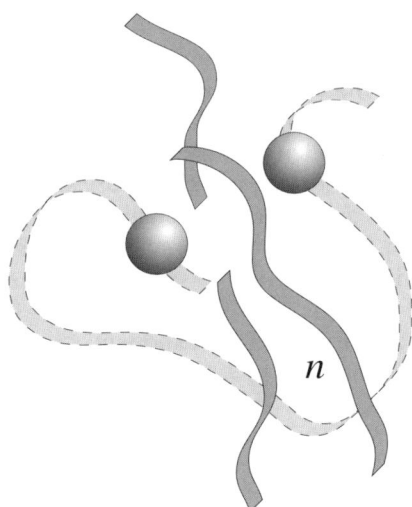

Fig. 3. Pair interaction is screened by other molecules

with n being the degree of polymerization of polymer molecules. Since we are seeking the osmotic pressure for polymers, we differentiate Eq. (22) with respect to v_p. And from the thermodynamic relation (V_p is the molar volume of polymers)

$$\Pi_p = -\frac{1}{V_p}\frac{\partial}{\partial v_p}\left(-T\Delta S + \Delta H\right)$$

one has

$$\Pi_p = RT\frac{c_s}{M_s}\left(1 + \left(\frac{1}{2n} - \chi\right)v_s + \cdots\right), \quad (24)$$

The first term in Eq. (24) represents the ideal term. If we consider an athermal system, it is sufficient to take into account the entropy mixing alone so that $\chi = 0$. And we have an important implication with the second virial coefficient A_2[1]:

$$A_2 \propto 1/n. \quad (25)$$

Like the Debye-Hückel shielding theory for the interaction between bare charges [27–29], the repulsion between monomers is effectively screened by other molecules, in proportion to the reciprocal of the molecular weight of surrounding materials (Fig. 3). Given the suggestion by Eq. (25), it appears reasonable to write the potential energy term of branched molecules corresponding to Eq. (19) in the more general form:

$$\frac{N^2}{r^{d+1}}\frac{1}{n}v \cong \frac{r}{\langle r_N^2 \rangle}. \quad (26)$$

For a monodisperse melt, $n = N$, whereas for the dilution limit, $n = 1$. In general, one may write $n = N^\varrho$, which leads to

$$r \propto N^v = N^{\frac{2(1+v_0)-\varrho}{d+2}}$$

whence

$$v = \frac{2(1+v_0)-\varrho}{d+2}, \quad (27)$$

---

[1] This n^{-1} dependence of the second virial coefficient has been already predicted in the Flory excluded volume theory [6]: let α be the expansion factor of a chain molecule and V_s the molar volume of solvents. Following Flory, one has

$$\alpha^5 - \alpha^3 = C'_M \frac{(1/2 - \chi)}{V_s} N^{\frac{1}{2}},$$

where C'_M is a coefficient. For a monodisperse melt, $V_s \propto n = N$ and $\chi = 0$. One has then $\alpha^5 - \alpha^3 \propto N^{-1/2}$, so that as $N \to \infty$, $\alpha \to 1$. In the absence of solvent, chains are nearly ideal.

where v_0 is the exponent for ideal molecules. Eq. (27) is the Isaacson-Lubensky result [30-32] based on the Flory concept and the screening idea. Eq. (27) neatly recovers the foregoing result

$$v = \frac{3}{d+2} \quad (v_0 = 1/2, \varrho = 0)$$

for a single linear chain in the dilution limit, and gives

$$v = \frac{3}{2(d+2)} \quad (v_0 = 1/4, \varrho = 1)$$

for a monodisperse branched melt.

Eq. (27) leads us to the aforementioned singularity concept, the marginal dimensionality, above which the excluded volume effects suddenly vanish: For instance, for a branched molecule in the dilution limit ($v_0 = 1/4, \varrho = 0$), Eq. (27) gives

$$v = \frac{5}{2(d+2)}$$

57which recovers the classical value, $v = 1/4$, for $d \geq 8$, thus $d_c = 8$ being suggested.

The Isaacson-Lubensky prediction [32] was confirmed by means of the enumeration of bond animals by Gaunt [33]. The shift of the excluded volume molecule to the ideal molecule is really a phase transition and not an asymptotic phenomenon [26, 30-35]. This aspect is again encountered in Sect. 6 and 7 in the context of the estimation of the gel point.

3.3
Numerical Estimation

Modern theories do not solve analytically the excluded volume problem, instead they pursue the same problem from different angles. This is because the excluded volume problem is placed as a category of typical many body problems [36], hence an intractable one. In spite of such a situation, there is an approach to persistently seek closed solutions of the excluded volume chain. Following this trend, some empirical formulations have been put forth for the limiting case of $N \to \infty$ [37-41]. The most successful one is that of the des Cloizeaux type equation, written by the form [41]:

$$p(|r|) \propto r^{2+\theta} \exp[-(Kr)^t], \quad \text{(for } N \to \infty\text{)} \tag{28}$$

with $|r|$ being an absolute length of the end-to-end distance rescaled as $|r|/\langle r_N^2 \rangle^{1/2} \to |r|$,

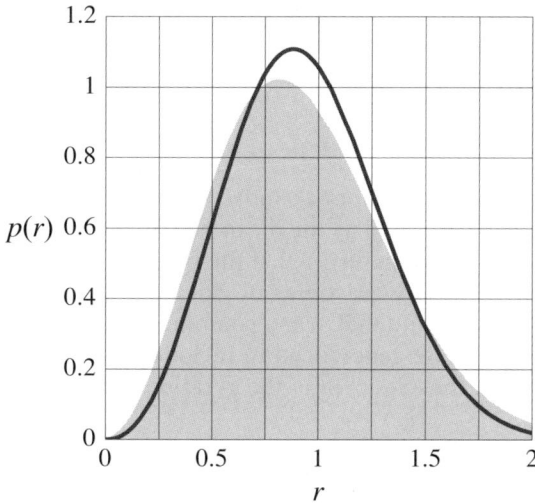

Fig. 4. Scaled end-to-end distance distribution for an excluded volume chain (solid line), Eq. (28), and the Gaussian chain (shaded area)

$$\theta = \frac{\gamma - 1}{\nu}, \text{ and } t = \frac{1}{1-\nu}.$$

ν and γ are the exponents of the mean end-to-end-distance $\langle r_N^2 \rangle \propto N^\nu$ and the number of self-avoiding walks $C_N \propto N^{\gamma-1}$, respectively. Because of the scale conversion, the normalization constant and K can be determined by way of the condition:

$$\int_0^\infty p(|r|)\,dr = \int_0^\infty r^2 p(|r|)\,dr = 1.$$

Using the best numerical values of $\nu = 0.588$ and $\gamma = 1.1619$ (hence, $\theta = 0.275$ and $t = 2.427$), and with the help of a technique of the pivot algorithm, Valleau [41] could show that the empirical equation remarkably well fits the Monte Carlo simulations on a $3d$-lattice, in support of the conjecture for a simple closed form of the excluded volume chain in the asymptotic limit, $N \to \infty$.

In Fig. 4, the conformational behavior of the excluded volume chain ($d = 3$) represented by Eq. (28) is drawn as a function of the rescaled end-to-end distance to be compared with that of the Gaussian chain. As one can see, the self-avoiding walk does not very much differ from the Gaussian statistics. In the vicinity of $r = 1$, the difference is only ≈10%. Thus, replacing real chains ($d = 3$) with the Gaussian chain is never a poor approximation. However, what is important is that the difference revealed in Fig. 4 between the excluded volume chain and the Gaussian chain does not disappear even in the limit of $N \to \infty$. In this real world ($d = 3$), one can never fit real chains to the Gaussian chain by properly altering the steric profile of structural units. The magical fit occurs above the critical dimension.

3.4
Conclusions

The essential argument of the Flory excluded volume theory is that pair-pair interactions can be well described by the mean potential energy, $\overline{U}(r)$, just like the van der Waals theory of real gases, so that the fraction of configuration reduces by the portion, $e^{-\overline{U}(r)/kT}$. This reasoning of Flory was remarkably successful for chain molecules. It was found that, when combined with the screening concept, the Flory argument can be generalized to include branched molecules, giving the general relation of the exponent of the gyration radius:

$$v = \frac{2(1+v_0) - \varrho}{d+2},$$

where v_0 is the exponent of the ideal molecule, d the space dimension, and $\varrho = 0$ for the dilution limit and $\varrho = 1$ for the melt. The equation predicts the existence of the critical dimension above which the ideal molecule emerges. In Flory's days, there was no concept of dimensionality, and no one was aware of this remarkable consequence. The profound implication of the Flory argument was discerned later by physicists.

4
Ring Formation

In this section, we derive ring distribution functions in the sol phase of real branching reactions to show how this result can be applied to the estimation of gel points in real systems. Although we push forward our argument taking a polycondensation as an example, the basic concept is, of course, applicable to all types of polymerizations as well.

A branching reaction proceeds either via an intermolecular or an intramolecular reaction. As far as the amount of rings is concerned, an important quantity is the relative velocity of cyclization to intermolecular reaction [42–46], and not the absolute ones.

Consider a chemical transition of a functional unit (FU) from an unreacted state to a reacted state, i.e., a single event of a bond formation. In place of the ordinary statistical description of the time transition, $t \rightarrow t + \Delta t$, we consider the bond transition, $i-1 \rightarrow i$ bonds [47]. This replacement is made simply because of physicochemical convenience. The expected fraction of a j-ring at this minute interval per unit bond formation ($\delta i = 1$) can be expressed as

$$p\{ring\ j\} = \frac{v_{R_j} - u_{R_j}}{v_L + v_R - u_L - u_R}, \tag{29}$$

where v_L denotes a total velocity of forward reactions for all intermolecular pairs and v_R for a total velocity of ring formation for all chains so that

$$v_R = \sum_{j=1}^{\infty} v_{R_j};$$

u_L and u_R are the corresponding velocities of the backward reactions. Now express v with the dimensions [$mole \cdot sec^{-1}$]; then the above probability can be equated with the mole fraction of j-rings to be formed during the minute interval $\delta i\, (=1)$. One can thus formulate [47]

$$\delta N_{R_j} = \frac{v_{R_j} - u_{R_j}}{v_L + v_R - u_L - u_R} \delta i. \tag{30}$$

Let M_0 be the total number of monomer units in the system. The extent of reaction, D, can be related to the total bond number i by the equality:

$$D = \frac{i}{\frac{1}{2} f M_0}, \quad \text{for} \quad R\text{-}A_f. \tag{31}$$

Substituting Eq. (31) into Eq. (30), and replacing the difference equation with a differential equation, one has

$$\left[N_{R_j}\right] = N_{R_j}/V \doteq \frac{1}{2} fC \int_D \frac{v_{R_j} - u_{R_j}}{v_L + v_R - u_L - u_R} dD, \tag{32}$$

with $C = M_0/V$. Eq. (32) is a fundamental equality of j-ring concentration. Now the problem to seek a ring distribution function in real polymer solutions has reduced to a problem to solve Eq. (32). Eq. (32) is a general expression that contains all information about ring formation in reversible and irreversible branching processes. Unfortunately, in general Eq. (32) can not be solved exactly; even for the most elementary rate equations of v and u, Eq. (32) does not yield an analytical solution. To date, only some asymptotic solutions has been found [47–49].

4.1
Cyclization in Real Systems

4.1.1
R-A_f Model

Consider a case where all backward reactions are null ($u = 0$) and the relative cyclization rate is negligibly small; this special system corresponds to an irreversible case in an infinitely concentrated solution ($C \to \infty$). Eq. (32) then reduces to

$$[N_{R_j}] = \frac{1}{2} fC \int_D (v_{R_j}/v_L) dD , \qquad (33)$$

The only task we should do is to estimate v_{R_j} and v_L. Let us confine ourselves to sufficiently slow reactions as compared with the diffusion process so that the mixing is so perfect as to excuse the approximation of quasi-static changes. Then, v_{R_j} is simply given by the product of the cyclization probability $\mathcal{P}$ of a j-chain and the total number of chances, ϕ_j, for j-ring formation

$$v_{R_j} = I \mathcal{P} \phi_j \exp(-\Delta H^\dagger / RT) , \qquad (34)$$

where I is a coefficient and the exponential term represents the probability due to the energy difference between the reactant and the transition state. $\mathcal{P}$ is the probability of one end of a j-chain entering a small volume v of the radius ℓ around the other end of the same chain, but is different from the ring closure probability P_{cy}. Quite generally, $\mathcal{P}$ can be written in the form:

$$\mathcal{P} = \int_0^\ell S_d P(r) dr , \qquad (35)$$

where S_d is the surface area of a d-dimensional sphere and $P(r) \cdot dr$ is the probability of the end-to-end distance of a j-chain lying between r and dr. For the Gaussian statistics, $\mathcal{P}$ can be reformulated in the form:

$$\mathcal{P} = 1 - \Gamma\left(\frac{d}{2}, \frac{d}{2v}\right) / \Gamma\left(\frac{d}{2}\right) , \qquad (35')$$

where we have used the relation, $\langle r_j^2 \rangle = v\ell^2$, and $\Gamma\left(\frac{d}{2}, \frac{d}{2v}\right)$ denotes an incomplete Gamma function defined by

$$\Gamma(a,z) = \int_z^\infty t^{a-1} e^{-t} dt .$$

To derive the total number of chances, ϕ_j, for j-ring formation, let us consider an m-tree of the R-A_f model which has m unreacted functional units (FU) in the root (Fig. 5).

Let $N(A)_j$ be the number of A functional units in the jth generation and D the fraction of reacted FUs. As one can see in Fig. 5, there are $f - m$ reacted FUs in the root (the first generation), $(f - m)(f - 1)$ FUs in the second generation, $(f - m)(f - 1)(f - 1)D$ FUs in the third generation, and so forth. it is easy to infer by induction that

$$N(A)_j = (f - m)(f - 1)[(f - 1)D]^{j-2} .$$

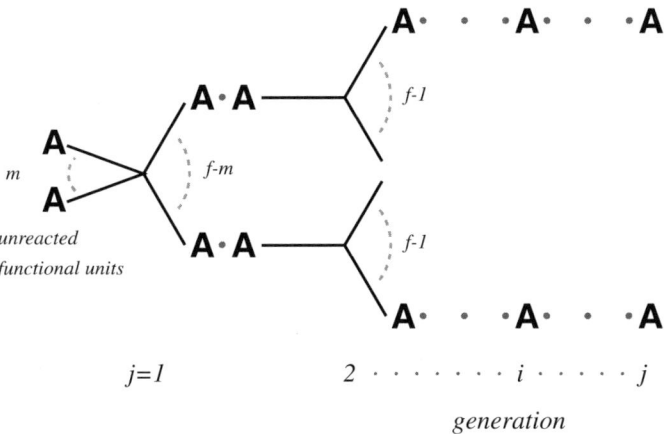

Fig. 5. A schematic representation of an m-tree that has m unreacted FUs in the root (R – A_f model)

Let M_i be the number of branching units that has i unreacted FUs. The distribution of $\{M_i\}$ is the binomial one. The number of branching units that has m unreacted FUs is therefore

$$M_0 \binom{f}{m}(1-D)^m D^{f-m}.$$

We can now express the number of chances of cyclization for j-chains:

$$\phi_j = \frac{1}{2} M_0 \sum_{m=0}^{f} m \times \binom{f}{m}(1-D)^m D^{f-m} N(A)_j (1-D),$$

which leads to

$$\phi_j = \frac{1}{2} f M_0 (f-1)(1-D)^2 \left[(f-1)D\right]^{j-1}. \tag{36}$$

With Eqs. (34) and (36), the rate of cyclization of j-chains can be written in the form:

$$v_{R_j} = \frac{1}{2} I \mathcal{P} f M_0 (f-1)(1-D)^2 \left[(f-1)D\right]^{j-1} \exp(-\Delta H^\dagger / RT). \tag{37}$$

Then let us proceed to formulate the rate of the intermolecular reaction. The probability of a given unreacted FU on a branched molecule entering a small region around a FU on the other molecule is $(v/V)\exp(-\Delta H^\dagger / RT)$, and there are $\frac{1}{2}[f M_0(1-D)]^2$ such intermolecular pairs. The rate of intermolecular reaction therefore becomes

$$v_L = \frac{1}{2} I [f M_0 (1-D)]^2 (v/V) \exp(-\Delta H^\dagger / RT), \left[\text{molecules} \cdot \sec^{-1} \right], \tag{38}$$

whence we have the asymptotic expression for the transition probability from $i-1$ to i bonds:

$$p\{ring\, j\} = v_{R_j}/v_L = \varphi_j (f-1)[(f-1)D]^{j-1} / fC, \quad (\text{for } C \to \infty), \tag{39}$$

where $\varphi_j = \mathcal{P}/v$ is the cyclization frequency of a j-chain relative to the intermolecular reaction, and $C = M_0/V$ denotes the initial monomer concentration. Substitute Eq. (39) into Eq. (33), then integrate with respect to D and we arrive at the solution of the ring distribution function in real polymer solutions:

$$[\Gamma]_{C \to \infty} = \sum_{j=1}^{\infty} \varphi_j [(f-1)D]^j / 2j, \quad (\text{for R-A}_f) \tag{40}$$

with $[\Gamma] = \sum_{j=1}^{\infty} [N_{R_j}]$. As one can see in the above derivation, Eq. (40) is valid for all space dimensions. This aspect of Eq. (40) is never trivial, as will be encountered in Sects. 6 and 7 in connection with the estimation of gel points.

4.1.2
R-A$_g$ + R-B$_{f-g}$ Model

In the same way, one has the solution for the R–A$_g$ + R–B$_{f-g}$ model:

$$[\Gamma]_{C \to \infty} = \sum_{j=1}^{\infty} \varphi_j [(g-1)(f-g-1)D_A D_B]^j / 2j, \tag{41}$$

where the subscripts A and B denote A and B FU, respectively.

4.1.3
A$_g$-R-B$_{f-g}$ Model

This model is somewhat different from the other models due to the existence of an extra factor ω_j and shows a characteristic behavior inherent to this model. So, the derivation is worth mentioning[47–49].

Consider again the m-tree which has m unreacted FUs in the first generation, and note a transit from the $(j-1)$th generation to the jth (see Fig. 6). As is seen from the illustration, a single A FU bears g A's and $f-g-1$ B's, while a single B FU bears $g-1$ As and $f-g$ Bs. The numbers of A FU and B FU in the jth generation then satisfy the recurrence relation of the form:

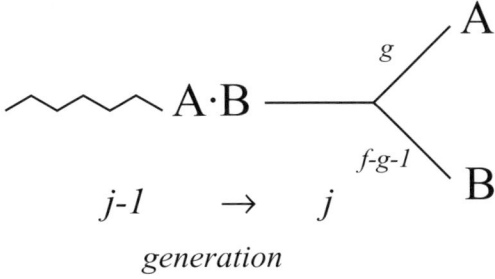

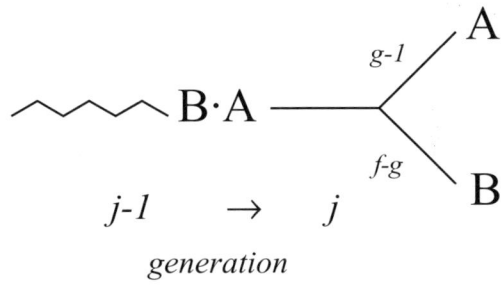

Fig. 6. A transit from the $(j-1)$th generation to the jth for the R–A$_g$+R–B$_{f-g}$ model

$$N(A)_j = gD_A N(A)_{j-1} + (g-1)D_B N(B)_{j-1};$$
$$N(B)_j = (f-g-1)D_A N(A)_{j-1} + (f-g)D_B N(B)_{j-1},$$

which can be rearranged in the matrix form:

$$\begin{pmatrix} N(A)_j \\ N(B)_j \end{pmatrix} = \begin{pmatrix} gD_A & (g-1)D_B \\ (f-g-1)D_A & (f-g)D_B \end{pmatrix} \begin{pmatrix} N(A)_{j-1} \\ N(B)_{j-1} \end{pmatrix}.$$

Now we write this in the form:

$$\tilde{N}_j = \tilde{A}\,\tilde{N}_{j-1},$$

which gives

$$\tilde{N}_j = \tilde{A}^{j-2}\,\tilde{N}_2, \qquad (42)$$

with $\tilde{N}_1 = \begin{pmatrix} g-m \\ f-g \end{pmatrix}$ and $\tilde{N}_2 = \begin{pmatrix} (g-m)g + (g-1)[(f-g)D_B] \\ (g-m)(f-g-1) + (f-g)[(f-g)D_B] \end{pmatrix}.$

The solution of Eq. (42) directly follows the Hamilton-Cayley theorem, the result being

$$\tilde{A}^k = \frac{e_1^k - e_2^k}{e_1 - e_2}\tilde{A} + \frac{e_1 e_2^k - e_1^k e_2}{e_1 - e_2}\tilde{E}. \qquad (43)$$

e_1 and e_2 are eigenvalues of the square matrix $\tilde{A}$ that satisfies

$$\det(\tilde{A} - \lambda\tilde{E}) = 0$$

where $\lambda = (e_1, e_2)$,
from which one finds

$$(e_1, e_2) = (f-g)D_B(1 \pm \alpha^{1/2}), \qquad (44)$$

where

$$\alpha = \frac{(g-1)(f-g-1)}{g(f-g)}.$$

From Eqs. (42)–(44), one has the expression of the numbers of FUs in the jth generation:

$$N(A)_j = (g-m)g[(f-g)D_B]^{j-2}\left(\frac{1}{2}\right)\left\{(1+\alpha^{1/2})^{j-1} + (1-\alpha^{1/2})^{j-1}\right\}$$

$$+ (g-1)[(f-g)D_B]^{j-1}\left(\frac{1}{2}\right)\alpha^{-1/2}\left\{(1+\alpha^{1/2})^{j-1} - (1-\alpha^{1/2})^{j-1}\right\};$$

$$N(B)_j = (g-m)(f-g-1)[(f-g)D_B]^{j-2}\left(\frac{1}{2}\right)\alpha^{-1/2}\left\{(1+\alpha^{1/2})^{j-1} - (1-\alpha^{1/2})^{j-1}\right\}$$

$$+ (f-g)[(f-g)D_B]^{j-1}\left(\frac{1}{2}\right)\left\{(1+\alpha^{1/2})^{j-1} + (1-\alpha^{1/2})^{j-1}\right\}. \qquad (45)$$

Note that $N(B)_j$ is the number of B FUs on a single m-tree, and the probability of finding an m-tree in the system again follows the binomial distribution. There exists M_0 such m-trees in the system. Hence, the total number of chances, ϕ_j, for j-ring formation can be calculated via

$$\phi_j = M_0 \sum_{m=0}^{g} m \times \binom{g}{m}(1-D_A)^m D_A^{g-m} N(B)_j (1-D_B),$$

which gives

$$\phi_j = M_0 g(1-D_A)(1-D_B)(f-g)[(f-g)D_B]^{j-1}\omega_j, \qquad (46)$$

where $\omega_j = \frac{1}{2}\left\{\left(1+\alpha^{1/2}\right)^j + \left(1-\alpha^{1/2}\right)^j\right\}$.

With Eq. (34), the cyclization rate can be expressed as

$$v_{R_j} = I\mathcal{P}M_0 g(1-D_A)(1-D_B)\left[(f-g)^j D_B^{j-1}\right]\omega_j \exp(-\Delta H^\dagger/RT),$$

while the intermolecular reaction rate is

$$v_L = I(v/V)M_0^2 g(f-g)(1-D_A)(1-D_B)\exp(-\Delta H^\dagger/RT).$$

Now substitute these rate equations into

$$[N_{R_j}] = gM_0 \int_D (v_{R_j}/v_L) dD_A, \tag{33'}$$

and one arrives at the asymptotic solution of the total ring concentration for the A_g-R-B_{f-g} model:

$$[\Gamma]_{C\to\infty} = \sum_{j=1}^{\infty} \varphi_j [gD_A]^j \omega_j/j$$

$$= \sum_{j=1}^{\infty} \varphi_j [(f-g)D_B]^j \omega_j/j, \quad \left(\text{for } A_g - R - B_{f-g}\right) \tag{47}$$

where ω_j is a factor specific to the A_g-R-B_{f-g} model which has the form:

$$\omega_j = \frac{1}{2}\left\{\left(1-\alpha^{1/2}\right)^j + \left(1+\alpha^{1/2}\right)^j\right\}; \alpha = \frac{(g-1)(f-g-1)}{g(f-g)}.$$

It is often useful to express Eq. (47) in another form:

$$[\Gamma]_{C\to\infty} = \sum_{j=1}^{\infty} \varphi_j \left[(f-g)(1+\alpha^{1/2})D_B\right]^j/2j +$$

$$\sum_{j=1}^{\infty} \varphi_j \left[(f-g)(1-\alpha^{1/2})D_B\right]^j/2j, \tag{47'}$$

Equally to the other model systems, the total ring concentration approaches a constant value, as $C\to\infty$.

Finding the Gel Point. Let us inspect how the gel point formula can be found from the above results. According to the definition, at the gel point an infinitely large cluster emerges, so that the total chances of j-ring formation must diverge,

$$\sum_{j=1}^{\infty} \phi_j \to \infty.$$

From Eq. (46) it follows that

$$\sum_{j=1}^{\infty} \phi_j = constant \cdot \left\{ \frac{1}{1-(f-g)D_B(1-\alpha^{1/2})} + \frac{1}{1-(f-g)D_B(1+\alpha^{1/2})} \right\} \to \infty$$

giving the gel point

$$D_{Bc} = \frac{1}{(f-g)(1\pm\alpha^{1/2})}. \tag{48}$$

These are the familiar classical gel points and have already been derived by Spouge [48] in a different context. The result of Eq. (48) is a quite natural consequence, since we are dealing with the hypothetical situation of an infinitely concentrated solution where the ring fraction is infinitely scarce so that the ideal tree model practically applies.

A curious feature of the A_g-R-B_{f-g} model is that it apparently possesses two critical points. But one of them is in fact physically unrealizable. Now consider an x-meric cluster of the A_g-R-B_{f-g} model without rings (we are considering the virtual tree molecule). The total number of B FUs within this molecule is $(f-g)x$, of which the number of reacted FUs amounts to $x-1$. The extent of reaction within this molecule, thus, may be written

$$D_B = \frac{x-1}{(f-g)x} = \frac{1-1/x}{f-g},$$

which increases monotonously from 0 to $\frac{1}{f-g}$, as x varies from 0 to ∞. And one has

$$D_B \leq \frac{1}{f-g}.$$

D_B can never exceed this upper bound, while by the boundary condition, $0 \leq \alpha < 1$, the aforementioned singularity formulae follows

$$0 \leq \frac{1}{(f-g)(1+\alpha^{1/2})} \leq \frac{1}{f-g} \leq \frac{1}{(f-g)(1-\alpha^{1/2})}. \tag{49}$$

The upper critical point can never be exceeded. We may abandon this physically unrealistic singularity, and thus

$$D_{BC} = \frac{1}{(f-g)(1+\alpha^{1/2})} \cdot (A_g - R - B_{f-g}) \tag{48'}$$

Through Sects. (4.1.1) to (4.1.3) we have derived the asymptotic solutions for the respective models. In spite of the limiting appearance of $C \to \infty$, these solutions (40), (41) and (47) are by no means fictitious entities devoid of practical importance, but have real meanings. They may be good approximations of $[\Gamma]$ in concentrated solutions such as non-solvent systems. This aspect of $[\Gamma]$ is discussed in some detail in the following since it is intimately connected with another important aspect of the gel point problem.

4.1.4
Physical Aspects of Asymptotic Equations

The assessment of the physical aspect of the foregoing solutions (40), (41) and (47) is a matter of critical importance, since in applying those equations to various problems, for instance, the distortion of the molecular weight distribution and the gel point shift, this process provides us with useful information.

At first sight, the above solutions may appear quite useless for practical purposes, because they are simply the limiting solutions of $C \to \infty$ where the relative frequency of cyclization is infinitely small so that w (weight fraction of rings) = 0. Contrary to our first impression of those equations, this is by no means the case. There is a definite evidence that the above asymptotic Eqs. (40), (41) and (47) are good approximations for ring concentrations in real processes.

For this discussion, an example of the bifunctional system ($f = 2$) is useful. As has been well-known [6], the bifunctional system has the general solution of the ring concentration written by a couple of equations of the form (the A – R – B model):

$$[\Gamma] = \sum_{x=1}^{\infty} \varphi_x \left(\frac{D-w}{1-w} \right)^x / x;$$

$$C = \frac{1}{w} \sum_{x=1}^{\infty} \varphi_x \left(\frac{D-w}{1-w} \right)^x. \tag{50}$$

In spite of the elementary mathematical form, the complete behavior of Eq. (50) has not been thoroughly investigated to date. Indeed, no quantitative analysis has been reported as to the C dependence of $[\Gamma]$. Thanks to the great advancement of modern computer science, however, we can now visualize the essential features of the multivariate Eq. (50). In Fig. 7, $[\Gamma]$ (molar concentration) and w (weight fraction) are plotted as a function of C for a fixed D (= 0.99), together with the corresponding asymptotic equation (broken line):

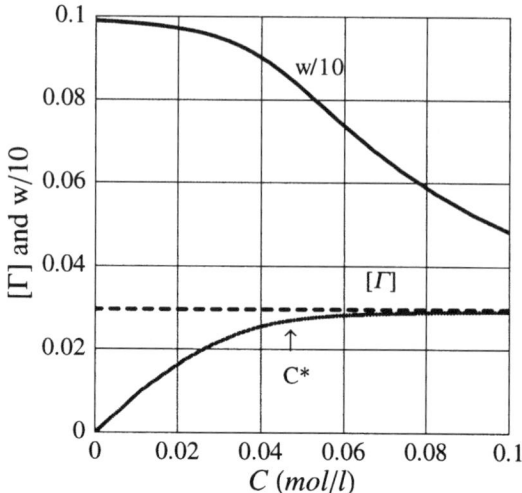

Fig. 7. An example of plot of Eq. (50): The molar concentration, $[\Gamma]$, and the weight fraction, w, of rings as a function of the initial monomer concentration, C ($D = 0.99$). The dotted line shows the asymptotic concentration:

$$[\Gamma]_{C \to \infty} = \sum_{x=1}^{\infty} \varphi_j D^j / j,$$

Eq. (50'), and C^* the boundary concentration beyond which the total ring concentration can approximately be equated with Eq. (50')

$$[\Gamma]_{C \to \infty} = \sum_{x=1}^{\infty} \varphi_j D^j / j, \quad (w \to 0) \tag{50'}$$

It is seen from Fig. 7 that there is a boundary concentration C^* below which the linear drop in $[\Gamma]$ occurs in parallel with the saturation of w (≥ 0.9; note that w cannot exceed D), showing that below C^* cyclization occurs predominantly over intermolecular reaction. Excess dilution beyond C^* leads to the linear drop of the net amount of rings.

More important is that above C^*, $[\Gamma]$ is $\approx constant$ which is nearly equal to the quantity predicted by the asymptotic Eq. (50') (dotted line). Now it turns out that for a given D the net amount of rings is invariable over a broad concentration range. This phenomenon has empirically been known among synthetic chemists, and qualitatively explained as follows:

If we let the system volume V increase by ΔV, the production of rings [$mole / l$] should decrease, because the total number of monomer units per unit volume (solute density) must decrease accordingly, whereas just by the same dilution effect, the relative cyclization rate to intermolecular reaction rate should increase, because cyclization follows the first order of the concentration, while the intermolecular reaction follows the second order, thus raising the yield of rings. These two actions work oppositely and cancel out each other, resulting in the observed concentration invariant of $[\Gamma]$.

We shall show below that the same concentration invariance can occur in general branching processes also.

From Eq. (32), it follows that for an irreversible case

$$[N_{R_j}] = \frac{1}{2} fC \int_D \frac{v_{R_j}/v_L}{1 + v_R/v_L} dD. \tag{51}$$

According to the concentration dependence of the rate equations, we may write Eq. (51) in the form:

$$[N_{R_j}] \approx \frac{1}{2} fC \int_D \frac{\varphi_j \mathcal{B}_j(D)/C}{1 + \sum_j^\infty \varphi_j \mathcal{B}_j(D)/C} dD, \tag{52}$$

where φ_j is the relative cyclization frequency defined in Sect. 4.1.1, and $\mathcal{B}_j(D)$ is a function of D, but independent of d and C. For a given D, if $C > \sum_j^\infty \varphi_j \mathcal{B}_j(D)$, we are lead to the foregoing solution (40), so that $[N_{R_j}] \cong constant$. Hence the quantity $\sum_j^\infty \varphi_j \mathcal{B}_j(D)$ gives a measure of the boundary concentration, $C^* \approx \sum_j^\infty \varphi_j \mathcal{B}_j(D)$.

Now it would be quite natural to consider that the same phenomenon observed in Fig. 7 occurs in general cases of $f > 2$; that is, for general irreversible processes, we may write

$$[\Gamma] = \sum_j^\infty [N_{R_j}] \cong constant \quad \text{for} \quad C > C^*, \tag{53}$$

for a given D.

In subsequent sections, we develop our arguments under the premise that Eq. (53) works as a general law. Since gelation is a phenomenon characteristic of concentrated solutions (most probably $C > C^*$), this premise greatly simplifies the theory.

4.1.5
Another Solution

There is another way to solve the fundamental equality (32). This method is more general, while incorporating many unknown parameters into the solutions. Still this approach is useful to outline cyclic production in branching media.

Consider again the irreversible process of $u = 0$. By Eq. (52), the total ring concentration should be given by

$$[\Gamma] \approx \frac{1}{2}fC\int_D \frac{\sum_j^\infty \varphi_j B_j(D)/C}{1+\sum_j^\infty \varphi_j B_j(D)/C}dD, \quad (R-A_f) \tag{54}$$

To solve this equation, we write Eq. (54) in a slightly more concrete form:

$$\sum_j^\infty \varphi_j \mathcal{B}_j(D)/C = \Theta(D)/fC.$$

Then Eq. (54) becomes

$$[\Gamma] \approx \frac{1}{2}fC\int_D \frac{\Theta(D)/fC}{1+\Theta(D)/fC}dD. \tag{55}$$

Now expand $\Theta(D)$ as a function of D to yield

$$\Theta(D) = \sum_{k=0}^\infty c_k D^k, \tag{56}$$

where c_js ($j=1,2,\cdots$) are unknown coefficients. Substituting Eq. (56) into Eq. (55), we have [47, 50]

$$[\Gamma] \approx \frac{1}{2}fC\int_D \frac{c_0 + c_1 D + c_2 D^2 + \cdots}{fC + c_0 + c_1 D + c_2 D^2 + \cdots}dD. \tag{57}$$

Let $\eta = fC + c_0$, and Eq. (57) reduces to

$$[\Gamma] \approx \frac{1}{2}fC\int_D \left[1 - \frac{fC/\eta}{1+(c_1/\eta)D+(c_2/\eta)D^2+\cdots}\right]dD$$

$$\equiv \frac{1}{2}fCD - \frac{1}{2}fC\int_D p\{inter\}dD. \tag{58}$$

In the limit of infinite dilution ($C \to 0$), $p\{inter\} \to 0$, so that cyclic production should lead to

$$[\Gamma]_{C \to 0} = \frac{1}{2}fCD. \quad \text{(for } C \to 0\text{)}$$

Thus the integral $\int_D p\{inter\}dD$ represents the decrement of cyclic production from $(1/2)fCD$. To solve Eq. (58), we neglect all the terms higher than the third [47]:

$$[\Gamma] \approx \frac{1}{2}fC\left\{D - \frac{fC}{c_2}\int_0^D \frac{1}{D^2+(c_1/c_2)D+\eta/c_2}dD\right\}, \tag{59}$$

and we have

$$[\Gamma] \approx \frac{1}{2} fC \left\{ D - \frac{2fC}{\sqrt{-c_1^2 + 4c_2\eta}} \left[\tan^{-1}\left(\frac{c_1 + 2c_2 D}{\sqrt{-c_1^2 + 4c_2\eta}} \right) - \tan^{-1}\left(\frac{c_1}{\sqrt{-c_1^2 + 4c_2\eta}} \right) \right] \right\}. \quad (60)$$

To visualize the general behavior of Eq. (60), let us assume the following correspondence,

$$c_k \leftrightarrow \varphi_{k+1} (f-1)^{k+1},$$

which can be inferred by taking the limit, $C \to \infty$, for Eq, (57) and comparing with Eq. (40). This treatment is not mathematically rigorous, but useful to outline the qualitative feature of Eq. (60). A numerical example of $[\Gamma]$ thus obtained is plotted in Fig. 8 ($f = 3$, $D = 0.25$, solid line), together with the asymptotic Eq. (40) (dotted line). As one can see, the same plateau in the high concentration regime as observed in the equilibrium bifunctional case ($f = 2$) is neatly reproduced. The small discrepancy between Eqs. (60) and (40) can be ascribed to the truncation of higher terms of c_k, which, however, suffices to convince us that the concentration invariant of $[\Gamma]$ is a general theorem.

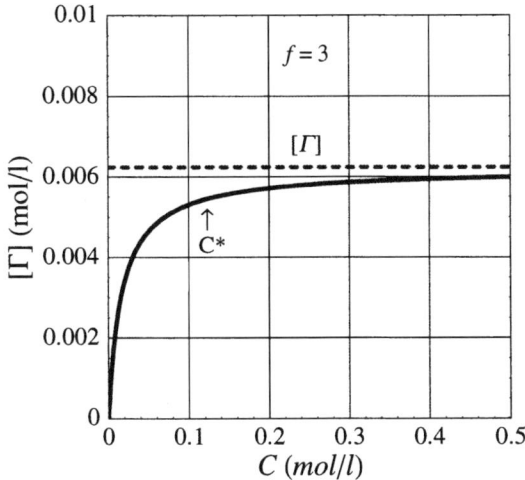

Fig. 8. An example of plot of Eq. (60) as a function of the initial monomer concentration, C ($D = 1/4$)

4.2
Cyclization on Lattice

4.2.1
Numerical Work

The mathematical task to analyze ring formation on a branched excluded volume molecule is a formidable one. To date, not a single general theory has been put forward to account for the branched molecule with rings; even for a single chain molecule, if the excluded volume effects must be taken into account, an exact solution cannot be obtained [51, 52]. For this reason, simulation experiments on lattices have often been utilized as the most convenient method, and various techniques have been devised [53, 54]. To visualize the random walk statistics, for instance, one first arranges the full length of a polymer molecule on lattices in an arbitrary configuration, then rotates one of bonds around the neighboring bond by a permitted angle (say, 90° for simple cubic lattices); the same procedure is followed successively over all combinations of all bonds and all permitted angles, thus realizing a self-avoiding walk. In this way the detailed study of the self avoiding walks became accessible. Such a simulation method was successfully applied, in a recent paper, to the study of the single polymer chain statistics as mentioned in Sect. 3.3 The merit of the computer simulation is obvious. It reduces an otherwise formidable mathematical problem to a simple problem of performance of a computer processor.

Cyclization of excluded volume chains was first investigated by Martin, Sykes, and Hioe [25, 55] on the triangular and face-centered cubic lattices. Counting a total number of self-avoiding walks, C_j, and the number, u_j, which return to the origin at the j th steps, they could show that the cyclization probability $P_{cy} = u_j / C_j$ varies asymptotically as $\sim j^{-11/6}$ for $d = 2$ and $\sim j^{-23/12}$ for $d = 3$ (these may be written approximately as $P_{cy} \sim j^{-2}$ for both $d = 2$ and 3). For simple random walks, the corresponding relations are $P_{cy} \sim j^{-1}$ and $\sim j^{-3/2}$, respectively, so that the excluded volume effects reduces the probability, P_{cy}, of return by $j^{5/6}$ for $d = 2$ and $j^{5/12}$ for $d = 3$, respectively (Fig. 9) [23, 25, 55–56]. It was suggested [25] that

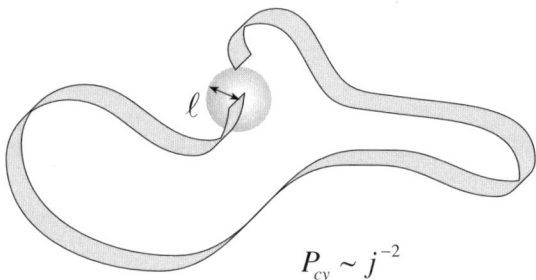

Fig. 9. A schematic representation of the probability, P_{cy}, of an excluded j-chain closing a ring

these results may have to be treated with caution, since a return to origin is a relatively rare event, and the behavior is anomalous at this point, so that they may not be representative of true P_{cy}.

An almost identical conclusion was obtained analytically by way of the self-consistent field theory. Edward [23] showed $P_{cy} \sim j^{-9/5}$ for $d = 3$ in good accord with the numerical calculation mentioned above, together with the mean end-to-end distance that scales as

$$\langle r_j^2 \rangle \propto j^{\frac{6}{d+2}},$$

giving the justification of the Flory excluded volume theory mentioned in Sect. 3.

4.2.2
Ring Distribution Functions on Lattices

A central concept for the derivation is again the transition probability [47, 57] introduced in Eq. (29). Within the framework of the deterministic description, the equalities of Eq. (29) and Eq. (30) is always valid for all chemical reactions, so that all one needs is to evaluate individual chemical reaction rates, say v_L and v_{R_j}. Let p_s be the fraction of sites to be occupied by monomer units. The possible bond number is then $\frac{1}{2} f M_0 p_s^2$. Considering that the typical percolation model is the R-A$_f$ type, and familiar branching reactions are irreversible, it suffices to solve the following equality:

$$[N_{R_j}] = \frac{1}{V} \int_D \frac{v_{R_j}}{v_L + v_R} di,$$

with $dD = di / \frac{1}{2} f M_0 p_s^2$, it follows that

$$[N_{R_j}] = \frac{1}{2} f C p_s^2 \int_D \frac{v_{R_j}/v_L}{1 + \sum_j^\infty v_{R_j}/v_L} dD. \tag{61}$$

The only one task we should perform is to find the relative rate v_{R_j}/v_L. Like the case in real system, we have not yet seen the complete descriptions of the rate equations, so that it would not be a realistic choice to attempt to work up a general solution of $[N_{R_j}]$. Thus, we again seek a limiting solution of Eq. (61).

Although, in Sect. 4.1, the limiting case of $C \to \infty$ was considered, the same thing is not applicable to the percolation process, since in the percolation model the concentration as a thermodynamic variable can not be defined in the ordinary chemical sense[57]. A similar situation, $p\{ring\ j\} \cong v_{R_j}/v_L$ $(v_L \gg v_R)$, can be realized in high dimensions [58]. In the following, we use this characteristic of the percolation model and derive an asymptotic solution of $d \to \infty$.

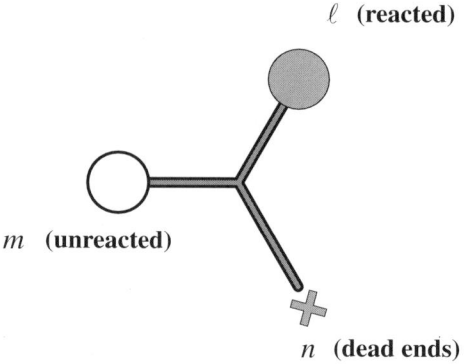

Fig. 10. A site on a lattice with m unreacted FUs, ℓ reacted FUs, and n dead ends

Cyclic Distribution in Bond Percolation Process. Consider a lattice with sufficiently large size and dimensions, $\mathscr{L}^d$ ($\mathscr{L} \to \infty$ and $d \to \infty$) so that clusters can be ideal. Following Eq. (61), we seek the analytical expression of the relative rate, v_{R_j}/v_L.

The cyclization rate v_{R_j} of j-chains is in proportion to the cyclization probability $\mathscr{P}$ multiplied by the total number ϕ_j of chances of cyclization of j-chains:

$$v_{R_j} \propto \mathscr{P} \times \phi_j. \tag{62}$$

$\mathscr{P}$ being the quantity defined in Eq. (35).

ϕ_j can be estimated as follows: Suppose an m-tree which has l reacted FUs, m unreacted FUs and n dead ends in the first generation (Fig. 10). Then, $l + m + n = f$, which follows the trinomial distribution [57, 84]:

$$\left[p_s D + p_s(1-D) + 1 - p_s\right]^f = \sum_{l=0}^{f} \sum_{m=0}^{f-l} \frac{f!}{l! \cdot m! \cdot n!} (p_s D)^l \{p_s(1-D)\}^m (1-p_s)^n. \tag{63}$$

Let $N(A)_j$ be the number of FUs in the jth generation. It is obvious that

$$N(A)_1 = l.$$

Each individual FU gives birth to new $(f-1)p_s$ FUs on average. So the number of FU in the 2nd generation is

$$N(A)_2 = l(f-1)p_s.$$

and

$$\begin{aligned} N(A)_j &= (f-1)p_s D \cdot N(A)_{j-1} \\ &= \left[(f-1)p_s D\right]^{j-2} N(A)_2. \end{aligned} \tag{64}$$

The total number of chances of cyclization for j-chains is calculated via

$$\phi_j = \frac{1}{2} M_0 p_s \sum_{l=0}^{f} \sum_{m=0}^{f-1} m \times \frac{f!}{l! \cdot m! \cdot n!} (p_s D)^l \{p_s(1-D)\}^m (1-p_s)^n N(A)_j (1-D), \quad (65)$$

which results in

$$\phi_j = \frac{1}{2} M_0 p_s^3 f(f-1)(1-D)^2 [(f-1)p_s D]^{j-1}. \quad (66)$$

Then we turn our attention to the self-avoiding walk for a lattice chain. Let, after j steps, one end enter the volume v around the other end; more exactly on anyone of the neighboring $f-1$ sites. Because of excluded volume effects, immediate reversals are forbidden for the end, so there exist $(f-1)p_s$ possible paths (FUs) for another step. The probability of these paths being vacant (unreacted) is $1-D$. The total number of paths available for the end is therefore $(f-1)p_s(1-D)$. Of these only one path can lead to the other end. Thus the probability of the two ends closing a ring can be written as

$$P_{cy} = \frac{\mathcal{P}}{(f-1)p_s(1-D)}, \quad (67)$$

whence the cyclization rate can be formulated as

$$v_{R_j} = I' \left\{ \frac{\mathcal{P}}{(f-1)p_s(1-D)} \right\} \left\{ \frac{1}{2} M_0 p_s^3 f(f-1)(1-D)^2 [(f-1)p_s D]^{j-1} \right\}, \quad (68)$$

where I' is another coefficient.

Then we proceed to the calculation of the intermolecular reaction rate. The intermolecular reaction rate v_L can be equated with the product of the probability of a given FU on a cluster entering into a small volume v with the radius of a bond length ℓ around a FU on the other cluster and the total number of all pairs (we put ℓ being the size of unit cell). Note that molecules are fixed on lattices, and each FU has a single chance to react, so it can react with the nearest neighbor alone, whereas the nearest neighbor is always unreacted because of the lattice specificity. Taking account of this specificity of the lattice model and that there is no heat of formation, one can write the intermolecular reaction rate as

$$v_L = \frac{1}{2} I' f M_0 p_s^2 (1-D) \times 1. \quad (69)$$

The expression asymptotically becomes exact as $d \to \infty$ where cyclization is suppressed entirely.

Comparing Eq. (69) with the corresponding expression (38) of real systems, it turns out that there are substantial differences between them: (i) the intermolecular reaction rate, v_L, of real systems follows the second order of M_0, while that of the percolation model the first order, and (ii) the entropy term, v/V, is absent from the latter reaction. These differences inevitably lead to differences in the cyclization mode, the dimensionality and the critical point behavior between them as will be seen subsequently.

From Eqs. (68) and (69), the relative rate becomes

$$v_{R_j}/v_L = \mathcal{P}\left[(f-1)p_s D\right]^{j-1}. \tag{70}$$

$\mathcal{P}$ is a monotonously decreasing function of d, so that the aforementioned transition probability asymptotically approaches

$$p\{ring\ j\} \to v_{R_j}/v_L, \quad \text{as} \quad d \to \infty. \tag{71}$$

And Eq. (61) reduces to

$$[\Gamma] = \sum_j^\infty [N_{R_j}] = \frac{1}{2}fCp_s^2 \sum_j^\infty \int_D \left(v_{R_j}/v_L\right) dD. \tag{72}$$

Substituting Eq. (70) into Eq. (72), one arrives at the asymptotic solution of the ring distribution function for the bond percolation problem:

$$[\Gamma]_{d\to\infty} = \frac{f}{(f-1)}Cp_s \sum_j^\infty 3\left[(f-1)p_s D\right]^j / 2j. \tag{73}$$

If f is an increasing function of d, the prefactor may be approximated as

$$\frac{f}{(f-1)} \cong 1, \quad \text{for large } f.$$

Then Eq. (73) reduces to

$$[\Gamma]_{d\to\infty} \cong Cp_s \sum_j^\infty 3\left[(f-1)p_s D\right]^j / 2j. \tag{73'}$$

It will be useful to rewrite Eq. (73') into the more familiar form. With the equality $C = (1/\ell)^d$ for the hypercubic lattices together with the definition $\varphi_j = \mathcal{P}/v$, one can reformulate Eq. (73') in the form:

$$[\Gamma]_{d\to\infty} \cong \#p_s \sum_j^\infty \varphi_j\left[(f-1)p_s D\right]^j / 2j, \tag{73''}$$

where $\mathscr{C} = 2\pi^{d/2}/d\Gamma\left(\frac{d}{2}\right)$ is a volume of the d-dimensional sphere with a radius of unity; that is, $v = \mathscr{C}\ell^d$.

It may be useful to inspect the critical point of the bond percolation model. As the critical point is approached, the total number, $\Sigma_j \phi_j$, of chances of cyclization should diverge. From Eq. (66), it follows that

$$\sum_j \phi_j \propto \frac{1}{1-(f-1)p_s D} \to \infty, \tag{74}$$

giving

$$D_c = \frac{1}{(f-1)p_s}. \tag{75}$$

This is the so-called ideal value on the Bethe lattice, which, when $p_s = 1$, recovers the Flory formula.

By analogy with real systems (Sect. 4.1), it seems natural to consider that Eq. (73) represents an approximate expression of $[\Gamma]$ in real percolation processes; that is, it would be natural to write for high dimensions

$$[\Gamma] = \sum_j^\infty [N_{R_j}] \cong constant \quad \text{for } d > d^*, \tag{76}$$

where d^* denoting a boundary dimensionality above which Eq. (76) works approximately. In Sect. 6, we will make full use of this relation to find a more general formula of the bond percolation threshold.

4.3
Conclusions

The molar concentration of rings in branching media is given by the general formula:

$$[\Gamma] = \sum_{j=1}^\infty \int_D^\infty \frac{v_{R_j}/v_L}{1+v_R/v_L} di, \quad \text{for irreversible reactions,}$$

where v_L is the intermolecular reaction rate, v_{R_j} the cyclization rate of j-chains and $v_R = \Sigma_{j=1}^\infty v_{R_j}$, and i the total number of bonds in the system. This is a fundamental equality that includes all information about cyclization phenomena of irreversible branching processes. The problem of seeking the cyclic distribution thus reduces to the problem of solving this fundamental equality. When we restrict ourselves to the limiting case of $C \to \infty$ or $d \to \infty$, this gives neat solutions: for real systems,
1. R–A_f model

$$[\Gamma]_{C \to \infty} = \sum_{j=1}^\infty \varphi_j [(f-1)D]^j / 2j; \tag{40}$$

2. $R\text{-}A_g + R\text{-}B_{f-g}$ model

$$[\Gamma]_{C\to\infty} = \sum_{j=1}^{\infty} \varphi_j \big[(g-1)(f-g-1)D_A D_B\big]^j / 2j. \tag{41}$$

3. $A_g\text{-}R\text{-}B_{f-g}$ model

$$[\Gamma]_{C\to\infty} = \sum_{j=1}^{\infty} \varphi_j \big[(f-g)D_B\big]^j \omega_j / j, \tag{47}$$

and for the

4. percolation model on hypercubic lattices

$$[\Gamma]_{d\to\infty} = \frac{f}{(f-1)} C p_s \sum_{j}^{\infty} 3\big[(f-1)p_s D\big]^j / 2j. \tag{73}$$

The last conclusion of this chapter is that these limiting solutions may be good approximations of $[\Gamma]$ under ordinary reaction conditions away from the limits. For instance, we may have for $C > C^*$

$$[\Gamma]_{C\to\infty} \cong [\Gamma], \tag{53}$$

as shown in Figs. (7) and (8).

5
Real Gelation and Percolation Process

In the course of the derivation of the ring distribution functions in branching media, it was found that analogies and differences coexist between real gelations and the percolation process [57, 84]. Let us summarize this aspect in perspective, since this procedure is essential to understand the differences in dimensional behaviors of the two systems which are encountered in Sects. 6 and 7. The respective ring distribution functions are of the form for the $R\text{-}A_f$ model:

$$[\Gamma]_{C\to\infty} = \sum_{j=1}^{\infty} \varphi_j \big[(f-1)D\big]^j / 2j, \quad (C\to\infty \text{ for real systems}); \tag{40}$$

$$[\Gamma]_{d\to\infty} \cong \# p_s \sum_{j}^{\infty} \varphi_j \big[(f-1)p_s D\big]^j / 2j. \quad (d\to\infty \text{ for the percolation model}) \tag{73''}$$

Although these are limiting expressions, one can expect them to work as good approximations of $[\Gamma]$ in actual processes if we are in high concentrations or at high dimensions, as discussed above.

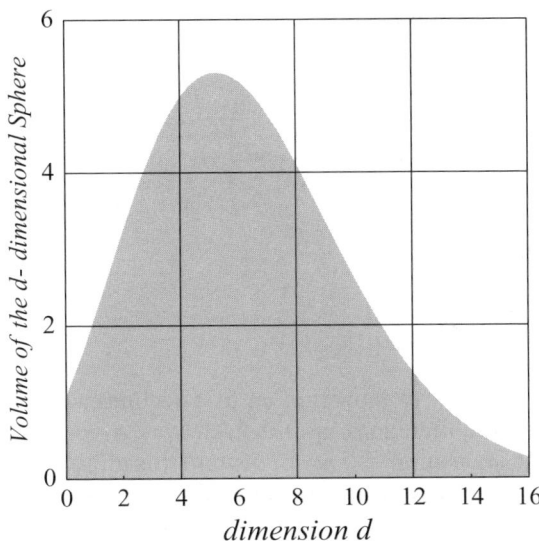

Fig. 11. The dimension dependence of the volume of the d-dimensional sphere with the radius of $\ell = 1$

For the pure bond case ($p_s = 1$), the two expressions are identical except for the coefficient $\mathscr{C} = 2\pi^{d/2}/d\,\Gamma(\frac{d}{2})$. Thus, it follows that whereas the microscopic feature of the chemical process is entirely different between real systems and the percolation model, the branching reaction proceeds, in effect, in a similar way. The only difference is the presence of the coefficient $\mathscr{C}$ that varies with dimensions as $4\pi/3$ ($d = 3$), $\pi^2/2$ ($d = 4$), …, having a maximum at $d = 5$, beyond which it rapidly decreases to 0 (Fig. 11). This does not mean that cyclization on lattices becomes maximum at $d = 5$. unlike $\mathscr{C}$ itself, the product $\mathscr{C} \times \varphi_j$ is a steeply decreasing function of d and so is the $[\Gamma]$, consistent with the observations in the percolation model.

If we confine our discussion to dimensions satisfying $\mathscr{C} \geq 1$ together with $d > d^*$ so that Eq. (76) is fulfilled, then it follows that the percolation model should produce rings excessively by the factor $\mathscr{C}$ than real systems do for $C > C^*$.

Major differences between Eqs. (40) and (73") can be summarized as follows:
(i) The ring fraction, $[\Gamma]/C$ ($p_s = 1$), on lattices is independent of C (note that $\mathscr{C}\varphi_j = C\mathscr{P}$), while that of real systems is inversely proportional to C.
(ii) More significant is the difference in dimensionality: The cyclization probability $\mathscr{P}$ is a decreasing function of d. By Eq. (73), the cyclic production on lattices must decrease with increasing dimensions (this was confirmed by the observations in the simulation experiments). In contrast, in real branching reactions it is the relative cyclization frequency φ_j that controls the amount of rings and not $\mathscr{P}$. Unlike $\mathscr{P}$, the quantity φ_j is not such a monotonous decreasing function of d. To see this, consider a hypothetical dense solution where the Gaussian

behavior is expected to apply to branched molecules. According to the relation, $\varphi_j = \mathcal{P}/v$, it follows that

$$\varphi_j = \frac{d}{\ell^d}\int_0^\ell r^{d-1}\left(\frac{d}{2\pi\langle r_j^2\rangle}\right)^{\frac{d}{2}}\exp\left(-\frac{d}{2\langle r_j^2\rangle}r^2\right)\cdot dr, \tag{77}$$

which can be equated with

$$\varphi_j = \frac{d}{2\pi^{d/2}\ell^d}\left[\Gamma\left(\frac{d}{2}\right)-\Gamma\left(\frac{d}{2},\frac{d}{2v}\right)\right], \tag{78}$$

where $\langle r_j^2\rangle = v\ell^2$. It is easy to show that φ_j first declines with increasing d to attain a minimum point, then goes up indefinitely as $d\to\infty$ (Fig. 12). There is a minimum point of φ_j around $d\approx\pi\langle r_j^2\rangle$. Beyond this minimum it turns to an increasing function of d. This means that at sufficiently high dimensions the system produces rings exclusively [57]. In Fig. 12 is shown an example of d-dependence of φ_j based on the numerical calculation of Eq. (78).

The above statement can be confirmed in a more general way. Consider a branching process in real systems. By the same argument as in Eq. (52), it follows that for an irreversible process

$$p\{ring\} = \sum_{j=1}^\infty p\{ring\ j\} \approx \frac{\sum_j^\infty \varphi_j \mathcal{B}_j/C}{1+\sum_j^\infty \varphi_j \mathcal{B}_j/C}, \text{ for real systems.} \tag{79}$$

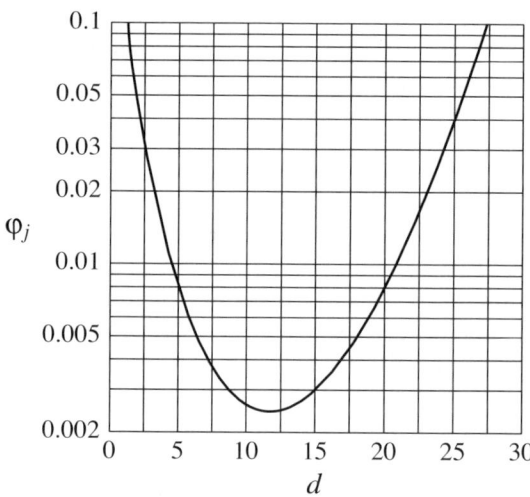

Fig. 12. The dimension dependence of the relative cyclization frequency, φ_j, for which Gaussian behavior is assumed

$\mathcal{B}_j$ is a function of D, but independent of the dimension. Let $\mathcal{N}$ be the number of monomer units on one dimensional length $\mathcal{L}$. Broadly speaking, for a large d, φ_j varies as $\sim d^{d/2}$, whereas $C = (\mathcal{N}/\mathcal{L})^d \sim constant^d$. φ_j is a stronger function of d than C itself, which leads Eq. (79) to $p\{ring\} \to 1$ and $p\{inter\} = 1 - p\{ring\} \to 0$, as $d \to \infty$. Hence, in real systems all the products should become cyclics as $d \to \infty$.

In light of the known dimension dependence of the percolation model, the conclusion drawn above will be a quite unexpected one, but the physicochemical reason is very clear without an ambiguity: As the dimensionality increases, the space-freedom should expand, so the probability of one end on a chain encountering the other end must decrease, while just by the same reason, the probability of two functional units on different molecules encountering must decrease as well. As a result of this competition, the minimum of φ_j should occur, beyond which cyclization becomes dominant over intermolecular reaction. At sufficiently high dimensions, cyclization finally overwhelms the intermolecular reaction, and the system behaves as if at a dilution limit.

In contrast to the situation in real systems, the collision probability on lattices between two units is always fixed to unity because of the structural specificity of lattices [19, 58–61]. Mathematically, this can be explained by the absence of the entropy term, v/V, in the intermolecular reaction on lattices. Consider an athermal reaction, and let N be the number of FUs. then the respective intermolecular reaction rates of Eqs. (38) and (69) can be simplified as follows

$$v_L = \tfrac{1}{2}N^2 \cdot (v/V) \equiv (\tfrac{1}{2}v) \cdot c \cdot N, \text{ for real systems;}$$

$$v_L = \tfrac{1}{2}N \cdot 1 \equiv \tfrac{1}{2} \cdot N, \text{ for the lattice model.}$$

At a glance, it can be understood that the intermolecular reaction on lattices has no dimension-dependence; it is a function of functionality f alone, while the cyclization rate, Eq. (68), depends strongly on d according to $\mathcal{P}$, the decreasing function of d. Thus, as the dimensionality increases, the cyclization rate alone naively declines, resulting in the known behavior of the percolation model: $p\{ring\} \to 0$ (Bethe lattice) as $d \to \infty$.

Through the comparative analysis of chemical machineries of branching processes between real systems and the percolation model, we have learned that the dimensionality is closely connected with cyclization. This feature does not appear to have been fully recognized by physicists up to the present. The reason simply comes from the fact that ring formation, together with excluded volume effects, has often been ignored, for the sake of mathematical simplicity, in the theory of branching processes. It is therefore not surprising that the unexpected dimension dependence of real systems remained unrealized so long.

5.1
Conclusions

The percolation model differs from real branching reactions in two points: (1) intermolecular reaction of the percolation model has no dimension dependence; (2) concentration in the ordinary chemical sense is absent in the percolation model. These differences arise from only the one fact that molecules are fixed on lattices, and give rise to the opposing dimensionality: as $d \to \infty$, in real systems cyclization becomes predominant, whereas in the percolation model it is suppressed entirely (the Bethe lattice). In these respects, the percolation model is not commensurate with the general features of the branching processes as chemical reactions.

6
Estimation of Gel Point

In a parallel with the theoretical advancement of gel research, a notable, new discovery was recently made in experimental polymer science. Tung and Dynes [62] could show that the viscoelastic method is a useful tool to determine the gel point. What is mentioned below is the essence of the Tung and Dynes' work.

6.1
Experimental Determination of Gel Point

For almost 50 years since the onset of the industrial manufacture of branched materials, the gel point has been estimated by measuring the solution viscosity combined with a solubility test of the resultant materials, which subsequently became the standard test of the gel point determination [63]. In 1982, an innovative experimental method was proposed. In their short paper, Tung and Dynes suggested a method that determines gel time from dynamic viscoelastic data. They put forward that the gel times of many thermosetting resins coincide with the crossover points of the dynamic storage G' and loss G'' moduli measured during isothermal setting; for instance, the cure behavior of an acetylene terminated sulfone (Fig. 13) showed that the intersection ($G' = G''$) of the two modulus curves occurs at $t = 59$ min., in correlation with the gel points as measured by the standard gelation test (Fig. 14).

The correlation between the crossover points and the gel points was remarkable, suggesting that the loss tangent ($G''/G' = \tan\delta$) is nearly unity at the gel point (Fig. 14). Tung and Dynes' explanation for this is as follows: the loss tangent, being the ratio of energy lost to energy stored in a cyclic deformation, measures the relative contribution of viscosity to elasticity of a resin system. The loss tangent of a viscous liquid therefore should be greater than unity, while that of an elastic solid should be less than unity. The resin systems proceed from a viscous liquid via gelation to an elastic solid. The loss tangent of the systems, corresponding to the transition state between viscous liquid and elastic solid,

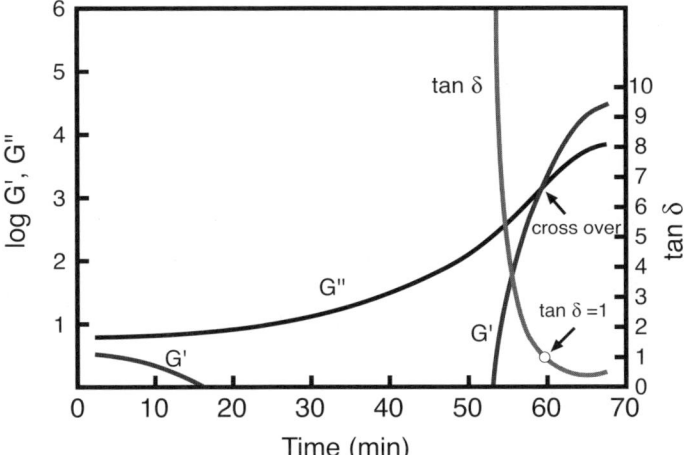

Fig. 13. Dynamic viscoelastic properties of ATS resin. Reproduced with permission from Journal of Applied Polymer Science, **27**, 571, 1982. Copyright 1982 John Wiley & Sons, Inc.

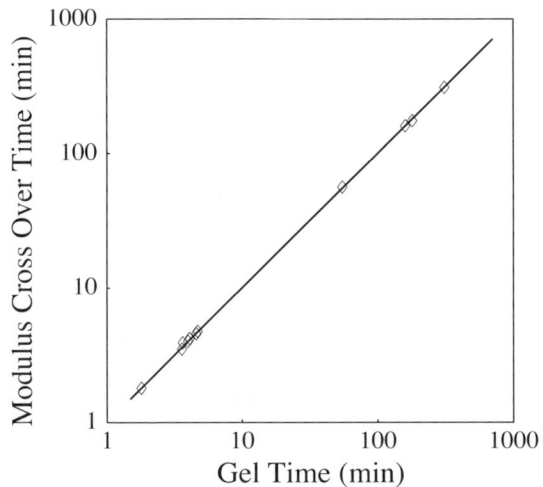

Fig. 14. Correlation of gel time and time of modulus crossover of thermosetting resins. Reproduced with permission from Journal of Applied Polymer Science, **27**, 572, 1982. Copyright 1982 John Wiley & Sons, Inc.

would therefore be expected to be equal to 1. Tung and Dynes' findings immediately spread into the community of polymer science, and spurred experimental activity. The relevance of their original idea was soon confirmed by many experimentalists [64–68]. Chambon and Winter carried Tung and Dynes' work further and put forward the empirical correlation between the moduli and applied frequency, ω, at the gel point:

$$\log G' / \log \omega = \log G'' / \log \omega \cong \frac{1}{2},$$

thus affirming the pertinence of Tung and Dynes' proposal over all frequency bands. In all the experiments, agreement between cross-over points of the moduli and gel points was remarkable. Tung and Dynes made a major breakthrough in experimental polymer science. This new finding therefore deserves to be called the "Tung and Dynes method" for their due credit. In Sect. 7, we will make use of the fruits of the Tung and Dynes method to test the theoretical equation.

6.2
Numerical Calculation

A characteristic feature of the gel point problem is that attempts to seek analytical solutions are surprisingly scarce. To date, only a few problems, the ideal tree model in polymer physics and two dimensional cases in the percolation model, have been solved analytically [6, 69–70]. No exact solutions including numerical solutions have been put forward for real systems with ring formation and excluded volume effects.

The reason becomes evident when one looks into the details of the gelation problem. There are too many intractable questions in this problem; the oldest, and most fundamental one is how one can enumerate all trees with different numbers and sizes of rings, and their entire combinations. In order to formulate the general solution, one must arrange infinite sets of (differential) equations, which is practically impossible. Hence, it is natural that the general consensus has arisen that there is no way to solve exactly the gel point problem.

For the percolation model, the situation is rather different, where the numerical calculation is easily accessible with the aid of the probability theory combined with a computational calculation. As a matter of course, the same methodology cannot be applied to real branching reactions, since real molecules are not fixed on lattices, and one cannot define the number of configurations that corresponds to the coordination number, q, of the percolation model.

Although, through the comparative analysis in Sect. 5, the percolation theory was found not to be equivalent to the branching process, we can still learn from this theory much deeper physics regarding the excluded volume problem, dimensionality, critical relations and many other important problems; in fact, we, polymer chemists, owe many startling discoveries of modern gel science to the percolation theory. In the following a physical method, now called the series expansion method is mentioned. This method was originated by Domb in 1961 as a promising approach for estimating the critical point (threshold) on lattices [25, 71–73]. Although the mathematical methods developed there are not directly connected with real gelations, they provide very useful clues when we think about the polymer gel point.

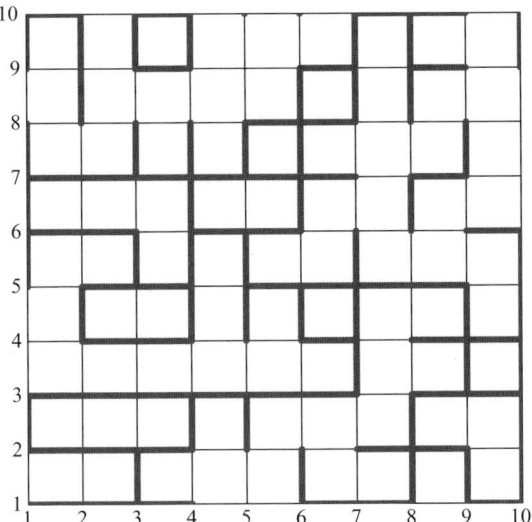

Fig. 15. An example of the bond percolation on a square lattice

Series Expansion Method (Percolation Model). Like the polymer theory [6], the percolation theory is based on the principle of the *equireactivity* of FU. In the percolation model, all mathematical complexity arising from ring formation and excluded volume effects is simply replaced with a set of random bonds. Let a site correspond to a monomer unit, then generate at random bonds between them, and we have an ensemble of bond animals (Fig. 15).

In order to estimate threshold values, let us introduce an additional assumption [25, 74]:
(i) The mean cluster size $S(D)$ can be expanded into the power series with positive coefficients a_j:

$$S(D) = \sum_{j=1}^{\infty} a_j \cdot D^j . \qquad (80)$$

This assumption seems well-founded, since the mean cluster size of the ideal tree model can also be expanded into a similar power series:

$$\langle x_w \rangle \propto \sum_{j=1}^{\infty} D_c^{-j} D^j \cdot \left(R - A_f \right) \qquad (81)$$

Given the assumption (i), according to the definition of the critical point, if $D < D_c$, the series of Eq. (80) should converge, while if $D \geq D_c$, it diverges. Clearly, the radius of convergence of the series corresponds to the threshold. Then, the d'Alembert convergence theorem [75] tells us

$$\varrho = \lim_{j\to\infty} \frac{a_j D_c^j}{a_{j+1} D_c^{j+1}} = \lim_{j\to\infty} \frac{a_j}{a_{j+1} D_c} = 1,$$

whence one gets the relation

$$\lim_{j\to\infty} \frac{a_j}{a_{j+1}} = D_c, \text{ (d'Alembert)} \qquad (82)$$

while the Cauchy theorem [75] provides another relation

$$\varrho = \lim_{j\to\infty} \sqrt[j]{a_j} \, D_c = 1,$$

and

$$\lim_{j\to\infty} 1/\sqrt[j]{a_j} = D_c. \text{ (Cauchy)} \qquad (83)$$

Now, all we need do is to evaluate the positive coefficient a_j on individual lattices. In principle, this can be done exactly by enumerating all configurations of given clusters. For instance, on the simple square lattice, one has $n_1(D) = (1-D)^4$ for $s = 1$; $n_2(D) = 2D(1-D)^6$ for $s = 2$; $n_3(D) = 6D^2(1-D)^8$ for $s = 3$; ... The terms consisting of $(1-D)^j$ are often called the perimeter polynomials and have been enumerated by Sykes and Essam up to the twelfth term [73]. Following Domb [71, 72], let us define the weight average cluster size $S(D)$ as

$$S(D) = \sum_{s=1}^{\infty} (s-1)^2 \cdot n_s(D) / 2D, \qquad (84)$$

namely, we count a bond as a monomer unit rather than a site. Rearranging the series in the ascending powers of D, we have

$$S(D) = 1 + 6D + 18D^2 + 48D^3 + 126D^4 + 300D^5 \\ + 762D^6 + 1668D^7 + 4216D^8 + 8668D^9 + 21988D^{10} + 43058D^{11} + \cdots, \qquad (85)$$

Taking the ratio of the 11th term to the 12th, we find

$$\frac{a_{11}}{a_{12}} = 0.5106\ldots,$$

the result being close to the exact value 1/2.

At first sight, it seemed obvious that the extrapolation to $j\to\infty$ intercepts the point $D_c = 0.5$ as predicted from the rigorous solution. Thus far, the series expansion method appeared very promising. Unexpectedly, the physicist's hope was suddenly dashed by a short note by Sykes and coworkers [76], the cofounders of the series expansion method. They made a detailed investigation for the

Table 1. Threshold values calculated from Eq. (86) and Monte Carlo simulation

d	2	3	4	5
Eq. (86)	0.7531	0.2502	0.1567	0.1166
Best Numerical values (Monte Carlo simulation)	0.5003	0.2470	0.1598	0.1181

configurations of larger clusters, adding several further coefficients to higher terms, and revealed that a weak singularity occurs for $D < D_c$; so the radius of convergence of the series cannot be identified with the threshold. The ratio method given in Eq. (82) suddenly lost its mathematical basis.

Gaunt and coworkers [77, 78] devised a more technical method. They expanded the coefficient a_j as a function of

$$D_{co}\left(=\frac{1}{2d-1}\right).$$

Making use of the Cauchy theorem, $\lim_{j\to\infty} 1/\sqrt[j]{a_j} = D_c$, they obtained for hypercubic lattices

$$D_c(bond) = \sigma^{-1}\left(1 + \tfrac{5}{2}\sigma^{-2} + \tfrac{15}{2}\sigma^{-3} + 57\sigma^{-4} + \cdots\right), \tag{86}$$

with $\sigma = 2d - 1$. Partly because of the truncation of higher terms, Eq. (86) slightly underestimates the threshold values for $d \geq 4$. In Table 1, the critical values calculated from Eq. (86) are compared with the best numerical values obtained by the Monte Carlo simulation [79, 80].

At present, the best numerical estimate of the threshold value is obtained through the Monte Carlo simulation on finite size lattices followed by the extrapolation to $\mathcal{L}$ (lattice size)$\to \infty$, the precision of which has been reported to be of the order 10^{-5} or higher.

6.3
Analytical Approach

6.3.1
Gel Point in Real Polymer Solutions

Let a branching unit possess f FUs. In order for an infinite molecule to appear, it is necessary that at least one functional unit (FU) have reacted; the probability of the remaining $f - 1$ FUs having reacted is $(f-1)D$. If $(f-1)D$ is greater than unity, then there is a finite probability that a given branching unit belongs to an infinite molecule. Thus the gel point should be given by

$$D_{co} = \frac{1}{f-1}, \tag{87}$$

with the subscript 0 representing the ideality of the classical gel point. The same conclusion can be drawn by way of many different derivations; e.g., via the percolation probability, $P(p)$, on Bethe lattices, the divergence of the number of chances, $\Sigma_j \phi_j$, of cyclization, and so forth, thus confirming the mathematical soundness of Eq. (87) as the emerging point of an infinite molecule. The experimental confirmation by Wile is of particular importance[48, 81, 82].

To improve on the original formulation of Eq. (87), a large amount of effort has been concentrated on taking ring structures into account. However, agreement with observations was unsatisfactory. In order to make a further advance in this field, another approach from a different angle has been needed. Although at present the Monte Carlo simulation enables us to identify thresholds in such high precision as mentioned above, such a numerical method does not provide us with physicochemical information about the critical point. Since the aim of gel science is not only to determine the location of D_c, but also to elucidate the physicochemical machinery of the gel point (threshold), it is of essential importance to understand why the gel point shifts from systems to systems. Hence, we must return to the original question of how the gel point is determined for branched systems with ring formation.

Throughout this and the next sections, we mention a theoretical method recently developed in the author's laboratory [57, 83–84]: First we derive the approximate expressions of the gel point for real systems, and then the corresponding expression for the percolation model. One will see there that the gel point is a unique point where the three effects, rings, excluded volume effects and dimensions, spontaneously appear and merge.

Then let us begin with the central concept of the theory.

Basic Concept. We confine ourselves to slow reactions so that the mixing of components is always perfect. Consider the R-A_f branching reaction that allows progressive bond-formation among A type FUs, where R represents a monomer unit and f the functionality. Given suitable reaction conditions, an infinite molecule can appear at a definitely defined point, the gel point. In general this point is mentioned in terms of the extent of reaction, D_c. It is important to notice that D_c is separable to the following two terms:

$$D_c = D(inter) + D(ring). \tag{88}$$

$D(inter)$ represents the extent of reaction of intermolecular reaction alone at the gel point, and $D(ring)$ that of cyclization alone. Note that only two FUs (one bond) are wasted every cyclization independently of size of rings [9]. Thus, it is convenient to define that cyclic bonds are equal to excess bonds which, when broken, do not disconnect a polymer molecule. Let $[\Gamma]$ be the number concentration (mol/l) of total rings and $C (= M_0/V)$ the initial number concentration of monomer units, and one has the equality:

$$D(ring) = \frac{2[\Gamma]}{fC} \quad (\text{R-A}_f). \tag{89}$$

Then let us proceed to the formulation of $D(inter)$. Suppose an equilibrium branching process where some fraction, p_R, of all the FUs, fM_0, is already occupied by cyclic bonds. The remaining FUs then form the equilibrium distribution, $\{f_i M_i\}$, where M_i denotes the number of monomer units having f_i FUs. Let us express the gel point of this mixing system in the form:

$$\frac{1}{\langle f \rangle - 1}, \quad (\text{R-A}_f)$$

which is equal to the ratio of the number of FUs consumed by intermolecular reaction as against the sum $\Sigma_i f_i M_i$ excluding the FUs wasted by cyclic bonds, whereas $D(inter)$ is the ratio against all FUs, fM_0. Following the definition of the extent of reaction, therefore, it is necessary to multiply the factor

$$\frac{\Sigma_i f_i M_i}{fM_0} \equiv 1 - \frac{Number\ of\ FU\ wasted\ by\ cyclization}{Total\ number\ of\ FU} = 1 - p_R.$$

p_R is the probability of a FU being occupied by cyclic bonds, and equivalent to $D(ring)$. Thus, $D(inter)$ can be expressed in the form:

$$D(inter) = (1 - p_R) \left\{ \frac{1}{\langle f \rangle - 1} \right\}. \tag{90}$$

Eqs. (89) and (90) lead Eq. (88) to the following analytical expression:

$$D_c = (1 - p_R) \left\{ \frac{1}{\langle f \rangle - 1} \right\} + \frac{2[\Gamma]}{fC}. \quad (\text{R-A}_f) \tag{91}$$

In the same way, for the R-A$_g$ + R-B$_{f\text{-}g}$ model where bond formation is permitted only between A type FU and B type FU, one has

$$D_c = (1 - p_R) \sqrt{\frac{1}{(\langle g \rangle - 1)(\langle f - g \rangle - 1)}} + \frac{[\Gamma]}{gC_A}. \quad (\text{R-A}_g + \text{R-B}_{f\text{-}g}) \tag{92}$$

for the equimolar case of the different functionalities, where C_A represents the initial monomer concentration for the A type monomer unit. Eqs. (91) and (92) are very general expressions that include all information about the gel point. Now the problem of estimating D_c has reduced to the problem of solving the above equalities.

Solving Basic Equalities. We introduce the following assumption.

Assumption I: cyclic bonds distribute randomly over all monomer units (*random distribution assumption of cyclic bonds*).

This assumption is equivalent to the statement that each FU has an equal chance to undergo cyclization. Given the random distribution assumption, the distribution of $\{M_i\}$ is then the binomial one. Now it is easy to find the mean functionalities $\langle \cdots \rangle$ which can be put equal to the weight average functionality defined by $\Sigma_i f_i^2 M_i / \Sigma_i f_i M_i$ [85–89]. Let $\overline{f_i^m}$ be the mth moment. The mean functionality is calculated via

$$\langle f \rangle = \frac{\overline{f_i^2}}{\overline{f_i}} = (f-1)(1-p_R)+1.$$

Substituting this result into the foregoing Eq. (91), one gets

$$D_c = \frac{1}{f-1} + \frac{2[\Gamma]}{fC}, \quad \text{for the R-A}_g + \text{R-B}_{f-g} \text{ model.} \tag{93}$$

and in the same way

$$D_c = \sqrt{\frac{1}{(g-1)(f-g-1)}} + \frac{[\Gamma]}{gC_A}, \quad \text{for the R-A}_g + \text{R-B}_{f-g} \text{ model.} \tag{94}$$

The problem of seeking analytic expressions of gel points has reduced to find the molar concentration of cyclics, $[\Gamma]$.

Finding General Solution of $[\Gamma]$. The fundamental equation for the ring distribution function is of the form [47] for the R-A$_f$ model:

$$[\Gamma]_{D=D_c} = \frac{fC}{2} \int_0^{D_c} \frac{\sum_j^\infty v_{R_j}/v_L}{1+\sum_j^\infty v_{R_j}/v_L} dD. \tag{95}$$

The integration extends over from 0 to D_c. Eq. (95) includes all information about the occurrence of rings in irreversible branching processes. Unfortunately, Eq. (95) is generally insolvable. Even for the most elementary rate equations, Eq. (95) does not yield an analytical expression. Thus we have to look for an approximate solution of Eq. (95). By the analysis in Sect. 4, it is known that, for high concentrations, $[\Gamma]$ is independent of C, but depends only on D and d, namely,

$$\left(\frac{\partial [\Gamma]}{\partial C}\right)_D \cong 0 \text{ (for high concentration), but } \left(\frac{\partial [\Gamma]}{\partial D}\right)_C > 0. \tag{96}$$

With these relationships in mind, we seek a solution of Eq. (95). Since $[\Gamma]$ is a continuous function of D_c, we can expand Eq. (95) with respect to $D_c = D_{co}$ which is realizable in the limit, $C \to \infty$; thus this expansion may be called a high concentration expansion. Designating $[\Gamma]$ as $Z(D_c)$, we find

$$Z(D_c) = Z(D_{co}) + \frac{Z'(D_{co})}{1!}(D_c - D_{co}) + \cdots. \tag{97}$$

Following the equality (91), D_c is also a function of the initial monomer concentration C, while as $C \to \infty$, $D_c \to D_{co}$ and $v_R/v_L \to 0$. Thus, the derivatives of Z can be written as

$$Z(D_{co}) = \frac{fC}{2} \sum_j \int_0^{D_{co}} (v_{R_j}/v_L) \cdot dD;$$

$$Z^{(1)}(D_{co}) = \left[\frac{fC}{2} \sum_j (v_{R_j}/v_L)\right]_{D_c = D_{co}}, \tag{98}$$

$$\cdots,$$

where (v_{R_j}/v_L) is a quantity at $C \to \infty$, and therefore can be replaced by Eq. (39):

$$v_{R_j}/v_L = \varphi_j(f-1)\left[(f-1)D_c\right]^{j-1}/fC, \quad \text{for } C \to \infty. \tag{39}$$

Substituting this into the above equalities and making use of the classical relation $(f-1)D_{co} = 1$, one gets

$$Z(D_{co}) = \frac{1}{2} \sum_j^\infty \varphi_j/j,$$

$$Z^{(1)}(D_{co}) = \frac{(f-1)}{2} \sum_j^\infty \varphi_j, \tag{99}$$

$$\cdots.$$

For a moment, let us suppose a special situation where a branched molecule is Gaussian so that $\varphi_j \propto j^{-\frac{d}{2}}$. Then it is obvious that the nth derivative, $Z^{(n)}$, diverges when $n \geq 1$ and $d \leq 2$. Thus, the expansion works only if the series is truncated to the first two terms together with $d \geq 3$. The manipulation is equivalent to applying the mean value theorem:

$$Z(D_c) = Z(D_{co}) + Z'(\xi)(D_c - D_{co}),$$

in which we are now making the approximation:

$$Z'(\xi) \to Z'(D_{co}).$$

Having given the mathematical framework of the expansion method, we collect the leading two terms of the series to get the approximate solution of $[\Gamma]$:

$$[\Gamma] \cong \sum_j \varphi_j / 2j + (f-1) \sum_j \varphi_j / 2 \cdot (D_c - D_{co}), \quad (d \geq 3). \tag{100}$$

Combining Eq. (93) with Eq. (100), after some rearrangement, we arrive at the analytical expression of the gel point ($d \geq 3$) for the
(a) R-A_f model

$$D_c = D_{co} \left\{ \frac{1 - \dfrac{f-1}{f} \sum_j (1 - 1/j) \varphi_j \cdot \gamma}{1 - \dfrac{f-1}{f} \sum_j \varphi_j \cdot \gamma} \right\}, \tag{101}$$

where $\gamma = 1/C$. In the same way, one has for the
(b) R-A_g + R-$B_{f\text{-}g}$ model

$$D_c = D_{co} \left\{ \frac{1 - \dfrac{f - g + g\kappa}{g(f-g)D_{co}} \sum_j (1 - 1/2j) \varphi_j \cdot \gamma}{1 - \dfrac{f - g + g\kappa}{g(f-g)D_{co}} \sum_j \varphi_j \cdot \gamma} \right\}, \tag{102}$$

where κ is the ratio of the number of FUs defined by $\kappa = (f-g) N_0 / g M_0 (\geq 1)$ with M_0 being the number of the A-type monomer units and N_0 that of the B-type units;

$$\gamma (=1/C) = \frac{V}{(M_0 + N_0)} = \left(\frac{f-g}{f-g+g\kappa} \right) \Big/ C_A;$$

D_c and D_{co} are the corresponding quantities for the A-type FU with

$$D_{co} = \sqrt{\frac{\kappa}{(g-1)(f-g-1)}}$$

[43–46, 90], and for the
(c) A_g-R-$B_{f\text{-}g}$ model

$$D_c = D_{co} \left\{ \frac{1 - (1/gD_{co}) \sum_j (1 - 1/j) \varphi_j [gD_{co}]^j \omega_j \cdot \gamma}{1 - (1/gD_{co}) \sum_j \varphi_j [gD_{co}]^j \omega_j \cdot \gamma} \right\}, \tag{103}$$

where $\omega_j = \frac{1}{2} \left\{ (1 - \alpha^{1/2})^j + (1 + \alpha^{1/2})^j \right\}$, $\alpha = (g-1)(f-g-1)/g(f-g)$,

$$D_{co} = \frac{1}{g(1+\alpha^{1/2})},$$ and D represents the quantity for the A FU likewise.

By virtue of the perturbation expansion with respect to $C \to \infty$, Eqs. (101)–(103) are exact in high concentrations. This is fortunate, because gelation is a phenomenon typical of concentrated systems [91, 92].

6.3.2
Critical Dilution

From the above solutions, the concept of a critical dilution, γ_c, naturally arises. For instance, using the boundary condition, $0 \leq D_c \leq 1$, one has for the R–A$_g$+R–B$_{f-g}$ model

$$\gamma_c = \frac{1-D_{co}}{\dfrac{f-g+g\kappa}{g(f-g)} \sum_{j=1}^{\infty}(1/D_{co} - 1 + 1/2j)\varphi_j}, \quad (\kappa \geq 1) \tag{104}$$

where D_{co} represents the quantity for the A-type FU as defined earlier. The gelation is possible only if we are below the critical dilution, γ_c [92], which reveals vividly that the gelation is a phenomenon typical of concentrated systems.

The notion of the critical dilution is in harmony with our intuition, suppose a branching process: molecular growth can take place only through the intermolecular linkages, whereas with increasing dilution the intermolecular reaction is suppressed. When the dilution reaches the well defined point, $\gamma_c = 1$, corresponding to $D_c = 1$, the probability of an infinite molecule emerging suddenly vanishes; hence, the critical dilution.

In the Flory model, no ring formation is allowed, and $\varphi_j \to 0$, so that from Eq. (104) it follows that $\gamma_c \to \infty$, thus the critical dilution disappears. In real gelations, such disappearance of γ_c never occurs because of the presence of the finite cyclization probability. The critical dilution is a general theorem in real gelations.

Reversible Gel. It may be useful to compare the critical dilution concept mentioned above with that of reversible gels. This phenomenon has been well-known in reversible gels [91]: we note that the forward reaction, the association of molecules (gelation), is bimolecular in nature, while the backward reaction, the dissociation, is unimolecular. With increasing dilution, therefore, the backward reaction should overcome the forward reaction, thus leading to the critical dilution beyond which gelation cannot occur. In this picture of the reversible gel, no cyclic formation is taken into account. The critical dilution in reversible gels can be explained within the framework of Le Chatelier's law, the analogue of Lenz's law which is more fundamental. Let us call an association point on chains a site, and consider a reversible reaction between unreacted sites and bonds. Let

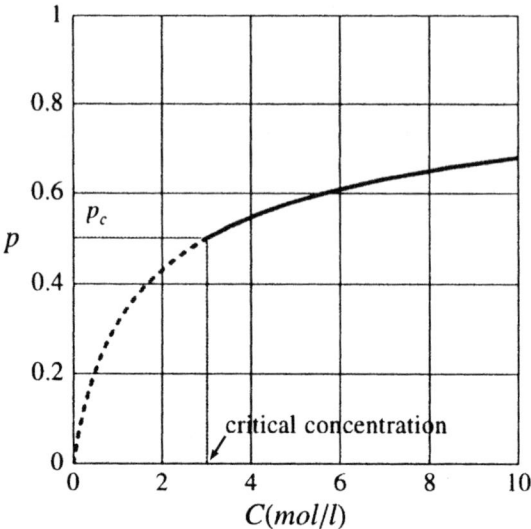

Fig. 16. A phase equilibrium in reversible gelation (K = 2/3). p_c is a critical value below which the bond probability can not exceed the classical value, $1/(f-1)$

p be the bond probability, $1-p$ the probability of unreacted sites, C the total site concentration, and K the equilibrium constant. By the law of mass action, one has the equilibrium condition:

$$K = \frac{1}{C}\frac{p}{(1-p)^2},$$

whence

$$C = \frac{1}{K}\frac{p}{(1-p)^2}.$$

As p changes from 0 to 1, C changes monotonously from 0 to infinity. Clearly there is a critical concentration below which p cannot exceed the critical value p_c, the gel point. Below this concentration, the bond probability can not exceed the well defined value $1/\langle f \rangle - 1$, the location of which is independent of dilution (Fig. 16).

6.3.3
Threshold in the Percolation Model

The percolation theory was introduced by Broadbent and Hammersley [93, 94] in 1957, 16 years after Flory's first paper [2] on gelation was published. It was nineteen years later when the theory was first applied to polymer chemistry as a realistic model of branching processes which takes into account ring formation and excluded volume effects [95, 96]. Since then, the percolation theory has developed in parallel with the advancement of gel science and made many predictions that have attracted so much attention in this field. The earliest and most familiar is the linear relationship [20, 97] between the percolation threshold and the coordination number z (functionality f). In 1961, Vyssotsky, Gordon, Frisch and Hammersley [97] observed that the threshold appears to be little affected by the difference of lattice types, but depends only on dimensions and coordination number z; they could show that there exists a hyperbolic relationship [20] for $d = 2$ and 3 between the percolation threshold and the coordination number z; that is,

$$z \cdot D_c \cong \frac{d}{d-1}.$$

Although the physical meaning of this correlation has not yet been fully understood to date [70], we expect that the present analysis would lead to a help for the deeper understanding [57].

In this section, we derive a theoretical equation for the percolation threshold using the high dimension expansion of $[\Gamma]$, and the result is examined in light of the corresponding Eq. (101) of real systems.

Theoretical. To find an approximate expression of the percolation threshold, it is necessary to make a slight modification of the fundamental equality Eq. (91). Note that the same as in real gelations, only one bond is wasted every ring formation, and the possible bond number is $\frac{1}{2} f M_0 p_s^2$, so that one can write

$$D_c = (1 - p_R) \left\{ \frac{1}{\langle f \rangle - 1} \right\} + \frac{2[\Gamma]}{fCp_s^2}, \quad \text{(site-bond problem)} \tag{105}$$

where p_R can be equated with $D(ring)$ so that $p_R = 2[\Gamma]/fCp_s^2$. If we accept the random distribution assumption of cyclic bonds (Assumption I), $\langle f \rangle$ can be equated with the weight average functionality. Then $\langle f \rangle$ is easily accessible using the moment generating function

$$M(\theta) = \left[p_s(1 - p_R)\theta + p_s p_R + 1 - p_s \right]^f, \tag{106}$$

namely,

$$\langle f \rangle = \frac{\overline{f^2}}{\overline{f}} = \frac{M'' + M'}{M'} \bigg|_{\theta=1} = (f-1)p_s(1 - p_R) + 1. \tag{107}$$

Substituting this result into Eq. (105), one finds

$$D_c = \frac{1}{(f-1)p_s} + \frac{2[\Gamma]}{fCp_s^2}, \tag{108}$$

the first term of the right hand side representing the ideal term (Bethe lattice) and the second term the correction term.

We proceed to estimate the ring concentration, $[\Gamma]$, on lattices. Using the lattice identity $dD = di / \frac{1}{2} f C p_s^2$ for $\mathscr{L} \to \infty$, we have the general expression of $[\Gamma]$:

$$[\Gamma]_{D=D_c} = \frac{f C p_s^2}{2} \int_0^{D_c} \frac{\sum_j^{\infty} v_{R_j}/v_L}{1 + \sum_j^{\infty} v_{R_j}/v_L} \cdot dD, \qquad (109)$$

By analogy with real systems, we assume the following inequalities that correspond to Eq. (96):

$$\left(\frac{\partial [\Gamma]}{\partial d}\right)_D \cong 0 \text{ (for high dimensions), but } \left(\frac{\partial [\Gamma]}{\partial D}\right)_d > 0. \qquad (110)$$

To find a solution of Eq. (109), we expand Eq. (109) with respect to $D_c = D_{co}$ which is realizable in the limit, $d \to \infty$. Thus this expansion may be called a high dimension expansion. In the same way as in real systems, one has

$$X(D_c) = X(D_{co}) + \frac{X'(D_{co})}{1!}(D_c - D_{co}) + \cdots, \qquad (111)$$

together with

$$X(D_{co}) = \frac{f C p_s^2}{2} \sum_j^{\infty} \int_0^{D_{co}} \left(v_{R_j}/v_L\right) \cdot dD,$$

$$X'(D_{co}) = \frac{f C p_s^2}{2} \sum_j^{\infty} \left(v_{R_j}/v_L\right), \qquad (112)$$

$$\cdots.$$

The expansion is valid for $d \geq 3$ by the same reason as stated for real systems. Since the limit, $D_c = D_{co}$, is attained with $1/d \to 0$, one can make use of the foregoing result, $v_{R_j}/v_L = \mathscr{P}[(f-1)p_s D]^{j-1}$ (Eq. (70)), derived in Sect. 4. Then using the relationship, $(f-1)p_s D_{co} = 1$ (Eq. (75)), one gets

$$X(D_c) = \frac{1}{2}\left(\frac{f}{f-1}\right) C p_s \sum_j^{\infty} \mathscr{P}/j + \frac{1}{2} f C p_s^2 (D_c - D_{co}) + \cdots. \qquad (113)$$

Collecting the leading two terms and substituting into Eq. (108), we arrive at the approximate solution for the percolation threshold[2]:

$$D_c = \frac{1}{(f-1)p_s}\left\{\frac{1-\sum_j^\infty (1-1/j)\mathcal{P}}{1-\sum_j^\infty \mathcal{P}}\right\}, \quad \text{(bond percolation)} \qquad (114)$$

for $d \geq 3$.

$\mathcal{P} \to 0$ as $d \to \infty$ is obvious, and one recovers Eq. (75) in the asymptotic limit. By virtue of the $1/d = 0$ expansion, Eq. (114) is exact in high dimensions. Eq. (114) states that the bond percolation threshold, D_c, should go up inversely in proportion to the site-occupation probability, p_s, in accord with the observations by Agrawal, Redner and Stanley [60], and by Stauffer [58].

It is believed that the site-occupation probability, p_s, plays a role of concentration in real systems. If that is the case, the hyperbolic relation between D_c and p_s predicted by Eq. (114) is in marked contrast to the corresponding D_c vs. C diagram of real systems, where D_c is known to converge to the ideal value, D_{co}, with $C \to \infty$ [48, 81, 86, 98, 99]. Clearly, the abnormality in the critical behavior of the percolation model is derived from the abnormality in the chemical machinery of the intermolecular reaction (Sect. 5) which is in proportion to the first order of C, more exactly to $M_0 p_s^2$, the same order as the cyclization rate, so that the concentration terms cancel out each other, resulting in the expression of Eq. (114) without the γ term.

6.4
Conclusions

The problem of estimating the location of gel points reduces to solving the following basic equalities:

$$D_c = (1-p_R)\sqrt{\frac{1}{(\langle g \rangle - 1)(\langle f-g \rangle - 1)}} + \frac{[\Gamma]}{gC_A}\left(R - A_g + R - B_{f-g}\right),$$

---

2 Identically, Eq. (114) can be deduced by means of the series expansion of Eq. (73). If the approximate expression (73') is used in place of Eq. (73), then the previous result [84] is recovered:

$$D_c = \frac{1}{(f-1)p_s}\left\{\frac{1-\frac{f-1}{f}\sum_j^\infty (1-1/j)\mathcal{P}}{1-\frac{f-1}{f}\sum_j^\infty \mathcal{P}}\right\}. \qquad (114')$$

The difference between Eq. (114) and Eq. (114') is technical rather than essential, but Eq. (114) slightly improves the agreement with the simulation values [58, 79] than Eq. (114').

for real systems, and

$$D_c = (1-p_R)\left\{\frac{1}{\langle f \rangle - 1}\right\} + \frac{2[\Gamma]}{fCp_s^2} \quad (\text{R-A}_f),$$

for the percolation model, respectively.

Under the assumption of the random distribution of cyclic bonds, and the expansion of $[\Gamma]$ about $D = D_{cc}$ followed by the truncation of higher terms, the above equalities yield the expressions of gel points for $d \geq 3$:

$$D_c = D_{co}\left\{\frac{1 - \frac{f-g+g\kappa}{g(f-g)D_{co}}\sum_j (1-1/2j)\varphi_j \cdot Y}{1 - \frac{f-g+g\kappa}{g(f-g)D_{co}}\sum_j \varphi_j \cdot Y}\right\} \quad (\text{R-A}_g + \text{R-B}_{f-g}), \qquad (102)$$

for real systems, and

$$D_c = \frac{1}{(f-1)p_s}\left\{\frac{1 - \sum_j^\infty (1-1/j)\mathcal{P}}{1 - \sum_j^\infty \mathcal{P}}\right\} \quad (\text{R-A}_f), \qquad (114)$$

for the percolation model, respectively.

By the boundary condition $0 \leq D_c \leq 1$, Eq. (102) gives the critical dilution beyond which gelation can not occur

$$Y_c = \frac{1 - D_{co}}{\frac{f-g+g\kappa}{g(f-g)}\sum_{j=1}^\infty (1/D_{co} - 1 + 1/2j)\varphi_j}. \quad (\kappa \geq 1) \qquad (104)$$

According to Eq. (104), the smaller κ and φ_j, the more Y_c should shift upward, and in the limit of $\varphi_j \to 0$ (tree model), the critical dilution vanishes.

7
Comparison with Experiments

7.1
Real Systems

The experimental reports on gel point determinations are scarce. The first reliable experiment was reported by Flory, who carried out this in 1941 in the polyesterification of a pentaerythritol - adipic acid mixture ($\text{R-A}_4 + \text{R-B}_2$) (Fig. 17) in order to test his own theory. In 1945, the gelation of the same system was reinvestigated in a more systematic way by Wile [81]; he observed that experimental gel points, when plotted against the reciprocal of the concentration, $\gamma\,(=1/C)$, are extrapolated exactly to the ideal value, D_{co}, with zero dilution, $\gamma=0$. About 20 years later, Gordon [98] reexamined in more detail the gelation of the same mixture to inspect the influence of the mole ratio, κ, on the concentration dependence of the gel point, and confirmed the linear extrapolation to the ideal value, D_{co}, as $\gamma\to0$. Ross-Murphy [99] investigated the concentration dependence of gel points in another polymer system, a benzene-1,3,5-triacetic acid-decamethyleneglycol mixture ($R-A_3+R-B_2$) diluted by 1,3,5-tris(diphenylmethylcarboxymethyl)benzene. He observed also that the same extrapolation to the ideal value occurs as $\gamma\to0$.

Relative Cyclization Frequency, φ_j. The above polyesterifications are of the $R-A_g+R-B_{f-g}$ type, and hence, we have to compare those observations with Eq. (102):

$$D_c = D_{co}\left\{\frac{1-\frac{f-g+g\kappa}{g(f-g)D_{co}}\sum_j(1-1/2j)\varphi_j\gamma}{1-\frac{f-g+g\kappa}{g(f-g)D_{co}}\sum_j\varphi_j\gamma}\right\}, \qquad (102)$$

where $\kappa=(f-g)N_0/gM_0$ (≥ 1), and D_c and D_{co} represent the quantities for the A-type FU and

$$D_{co}=\sqrt{\frac{\kappa}{(g-1)(f-g-1)}},$$

as defined earlier.

Only one unknown parameter with Eq. (102) is the relative cyclization frequency, $\varphi_j = \mathscr{P}/v$. Note that Eq. (102) was derived by means of the perturbation expansion with respect to $C\to\infty$ ($\gamma=0$). In this hypothetical limit, the cyclic production becomes negligible, and all sorts of excluded volume effects should vanish rigorously, thus realizing the ideal situation[3].

---

[3] Short-range properties, of course, still remain invariable as $C\to\infty$. To see this, consider a situation in which C increases from the dilution limit up to the melt, C_{melt}. no bond angles, no bond lengths and no rotational constraints change on average. Then, suppose a change from C_{melt} to C_∞. Mean effective radii of atoms (see Sect. 3) must decrease accordingly: The more the atoms are densely packed, the more the volume contraction becomes pronounced. In that extension does lie the hypothetical limit of C_∞, where all sorts of excluded volume effects should vanish rigorously.
When generalizing this reasoning to include ring formation, we may expect that the ideal tree model with no excluded volume and no ring formation will be realized in the limit C_∞. The classical gel theory corresponds to this limiting case ($C\to\infty$). That is thus comparable to the classic status of the ideal gas law ($C\to 0$) to the real gas.

While a chain within a branched molecule is by no means Gaussian in this real world ($d = 3$; note that we are considering the end-to-end distance distribution, and not the segment-density distribution about the center of gravity), by virtue of the perturbation expansion with respect to $C \to \infty$, one can apply the Gaussian approximation to φ_j. Then the most reliable way to estimate the unperturbed φ_j will be to consult the ring-chain equilibria in linear melts where situations are ideal [6, 30]. Prior to the evaluation of φ_j, it will be useful to review the thermodynamic relation[6, 102] between φ_j and the equilibrium constant, K.

Thermodynamic Equality. Consider the chemical equilibria between chains and an x-ring:

$$C_{x+y} \rightleftarrows C_x + C_y, \tag{c1}$$

$$C_x \rightleftarrows R_x, \tag{c2}$$

where x represents the number of monomer units. With the thermodynamic relation, it follows that

$$\Delta F = \Delta H - T\Delta S = -RT \ln K_x, \tag{c3}$$

whence for the equilibrium (c1)

$$K_1 = \frac{V}{2v} \exp(-\Delta H_1 / RT), \tag{c4}$$

and for (c2)

$$K_2 = \frac{1}{x} \int_0^\ell S_d \, P(r) \, dr \exp(-\Delta H_2 / RT), \tag{c5}$$

together with

$$K_x = K_1 K_2.$$

For large rings, the enthalpy (bond energy) terms should cancel out

$$\Delta H_1 + \Delta H_2 = 0,$$

and by definition,

$$\varphi_x = \mathscr{P}/v = (1/v)\int_0^\ell S_d \, P(r) \, dr, \tag{c6}$$

where S_d is the surface area of a d-dimensional sphere, $P(r)$ the end-to-end distance distribution of the x-chain, and ℓ the bond length. From the above results, and noting that $V = 1[l]$, one arrives at the thermodynamic equality:

$$K_x = \varphi_x / 2x. \quad (c7)$$

Inspired by the formulation of Eq. (c7), numerous experiments have been performed. Assuming the approximate equality, $K_x \approx [R_x]$ (molar concentration of x-rings), Jones and coworkers [100] measured the K_x values up to $x = 7$ in ring-chain equilibria of linear aliphatic polyesters. In Table 2 are summarized their results for $\Sigma_x \varphi_x$ and $\Sigma_x \varphi_x / 2x$, and the ratio, $r = \Sigma_x \varphi_x / \Sigma_x (\varphi_x / 2x)$, calculated via Eq. (c7), together with an example of the aromatic polyester[101–102], and poly(ethylene terephthalate) for reference.

As one can see in Table 2, the ratios, $r \approx 8$, are about twice as large as the values predicted from the power law relationship, $\Sigma_{x=1}^{\infty} x^{-3/2} / \Sigma_{x=1}^{\infty} (x^{-5/2}/2) \approx 4$, which, however, can be ascribed to less production of the smallest ring of the size $x = 1$. Note that $x^{-5/2}$ is a more rapidly decreasing function of x than $x^{-3/2}$, and the exclusion of the smallest ring directly raises the ratio, for instance, to $r = \Sigma_{x=2}^{\infty} x^{-3/2} / \Sigma_{x=2}^{\infty} (x^{-5/2}/2) \approx 9$, consistent with the experimental values in Table 2. For correct evaluation of φ_x, this ratio problem is a matter of critical importance. So let us inspect the chemical reason of this abnormality of r values more closely.

7.1.1
gem-Substituent Effects

The polyesters shown in Table 2 have repeating units consisting of 9 and 18 skeletal bonds, respectively, while the adipic acid-pentaerithritol polymer has a repeating unit of 11 bonds on the same sort of aliphatic backbone, an intermediate length of the polyesters shown in Table 2. Hence, Jones and coworkers' experiment is a good measure for the present estimate. An only difference between

Table 2. Observed relative cyclization frequencies of linear polyesters.

Polyester	$\sum_{x=1}^{\infty} \varphi_x / 2x$	$\sum_{x=1}^{\infty} \varphi_x$	r
Jones and coworkers[a]			
[O(CH$_2$)$_3$OCO(CH$_2$)$_2$CO]	0.100	0.830	8.3
[O(CH$_2$)$_{10}$OCO(CH$_2$)$_4$CO]	0.066	0.488	7.4
Cooper and Semlyen[b]			
[O(CH$_2$)$_2$OCO · C$_6$H$_4$ · CO]	(0.06)	(0.75)	(12.5)

Calculated, using the experimental values for small cyclics, and the Gaussian power law, $\varphi_x \propto x^{-3/2}$, for larger cyclics.
[a] From ref. [100].
[b] From ref. [101].

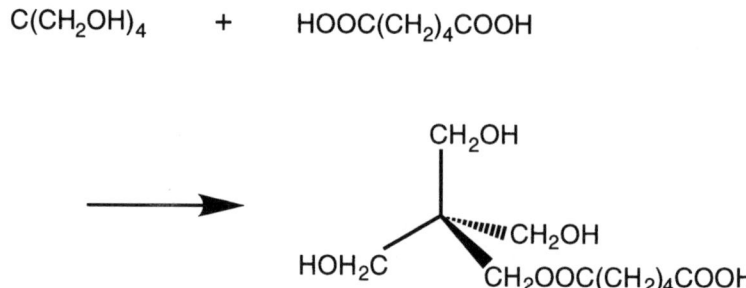

Fig. 17. A polycondensation reaction between pentaerithritol and adipic acid

those polyesters and the branched polyester (adipic acid – pentaerithritol polymer) in question is that the latter polyester has, by nature, the large *gem*-substituents, the hydroxy methyl moieties or their ester derivatives (see Fig. 17).

Early in the last century, it had been perceived by organic chemists that some sort of substituents greatly accelerate the rate of cyclization [103, 104]:

Azelaonitrile [N≡C-$(CH_2)_7$-C≡N] converts to cyclic ketone at about 30% yield according to the Thorpe-Ziegler reaction, while 5-t-butyl- and α,α'-dimethyl-azelaonitriles [N≡C-$C(CH_3)_2$-$(CH_2)_6$-C≡N] convert to the corresponding cyclic ketone at 89% and nearly exclusively, respectively, under the same reaction condition.

This type of substituent effect has been called "*gem*-dimethyl effect" or "Thorpe-Ingold effect" after Ingold and Thorpe [103], which has been interpreted as follows: In order for one end FU on a chain molecule to encounter the other end for cyclization, it is essential for the chain molecule to take the *gauche* conformation, requiring appreciable steric energy, on one hand. Since this steric effect, together with the transannular effects, appears the most remarkably in 8- to 20-membered ring formation, it has become called the "medium size ring effect", which in most cases depresses extremely the production of rings. The first indication of this steric effect appeared in Ruzicka's work [105] in 1920s on the preparation of cyclic ketones: He observed, that the yields of rings fell sharply to a very low minimum at the nine- to eleven-membered rings, and then rose slowly towards the still larger rings. The polyesters shown in Table 2 consist of this "medium length" repeating unit (9~18-ring). The foregoing ratio, r, abnormality (Table 2) can reasonably be attributed to the "medium size ring effect" that induces the depression of ring 1-mers [102–105].

Given *gem*-substituents, on the other hand, the energy of the *gauche* conformation to the *trans* conformation relatively vanishes, thus raising the relative frequency of cyclization to intermolecular reaction (Figs. 17 and 18).

Now that the constraint arising from the steric energy has been removed, one may assume, as a first approximation, that all the ring formations (from $x = 1$ to ∞) occur according to the known power law, $\varphi_x \propto x^{-3/2}$. Making use of the mean end-to-end distances computed from the rotational isomeric state models, one has

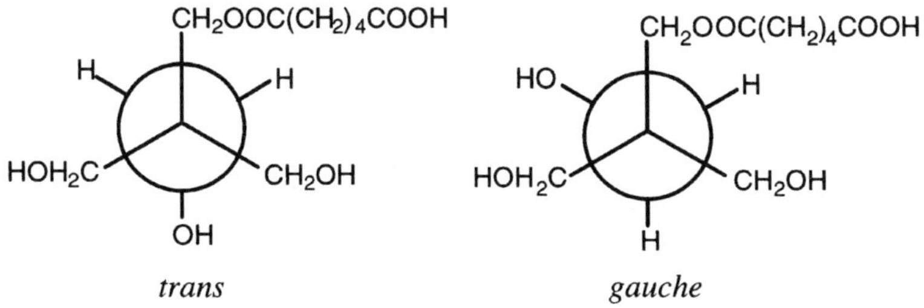

Fig. 18. Newman projection for the pentaerithritol-adipic acid dimer unit

$$\sum_{j=1}^{\infty} \varphi_j / 2j \approx 0.4, \quad \text{for} \quad \xi = 9,$$

and

$$\sum_{j=1}^{\infty} \varphi_j / 2j \approx 0.13, \quad \text{for} \quad \xi = 18,$$

where ξ is the number of skeletal bonds per repeating unit. Using these values and taking into account the repeating length, $\xi = 11$, of the polyester in question (Fig. 17), one has

$$\sum_{j=1}^{\infty} \varphi_j / 2j \approx 0.32 \ [mol \cdot l^{\pm 1}], \quad \text{and} \quad \sum_{j=1}^{\infty} \varphi_j \approx 4 \times 0.32. \tag{115}$$

Having determined the φ_j values, it is easy to compute, according to Eq. (102), the gel point, D_c. The result is plotted in Fig. 19 as a function of y, together with the Flory (△), the Wile (◇), and the Gordon (×) observations. The symbol (○) denotes the ideal point, D_{co}, and (⊗) the critical dilution, y_c.

Agreement between the theory and the experiments is excellent for all the regimes, in support of the mathematical soundness of the theory. By Eq. (104), it follows that $y_c = 0.448$. It is noteworthy that y_c is located in a very high concentration regime of $C > 1.0 \ mol/l$.

7.1.2
Without gem-Substituent Effects

To confirm the above findings, it will be essential to examine the theory with a branched polymer system without the *gem*-substituents effect. An experiment that meets this requirement has been reported by Ross-Murphy for the polycondensation system of 1,3,5-tris(carboxymethyl)benzene and decamethyleneglycol (Fig. 20).

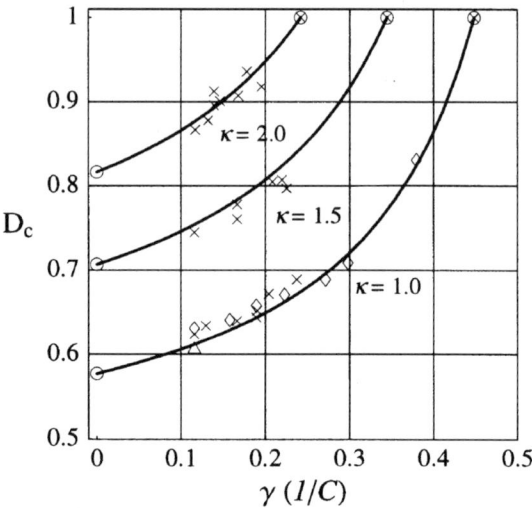

Fig. 19. Comparison of theory and experiment: D_c vs. $\gamma\,[l/mol]$ curve Solid lines (—): Eq. (102). (○): classical gel points; (⊗): critical dilution. Experimental points: by Flory (△: κ = 1), Wile (◇: κ = 1), and Gordon (×: κ = 1, 1.5 and 2.0), where

$$\kappa = \frac{[\text{COOH}]}{[\text{OH}]}$$

and D_c represents the extent of reaction of the A type functional units (OH)

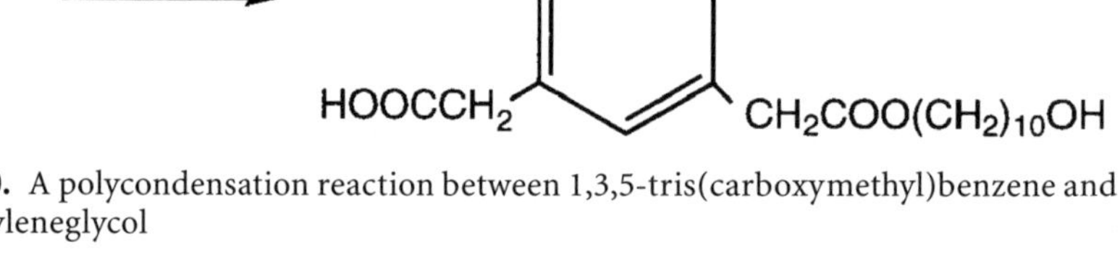

$$C_6H_3(CH_2COOH)_3 \quad + \quad HO(CH_2)_{10}OH$$

$$\longrightarrow$$

[structure: benzene ring with CH₂COOH, HOOCCH₂, and CH₂COO(CH₂)₁₀OH substituents]

Fig. 20. A polycondensation reaction between 1,3,5-tris(carboxymethyl)benzene and decamethyleneglycol

The structural unit of this branched ester is shown in Fig. 20. Because of the absence of *gem*-substituents, we expect the ordinary φ_j values for this polymer system. Then the work of Jones and coworkers is again a good measure. Taking the average of the observed values in Table 2, one has:

$$\sum_j \varphi_j / 2j = 0.085 \left[mol \cdot l^{-1} \right], \quad \text{and} \quad \sum_j \varphi_j = 8 \times 0.085. \tag{116}$$

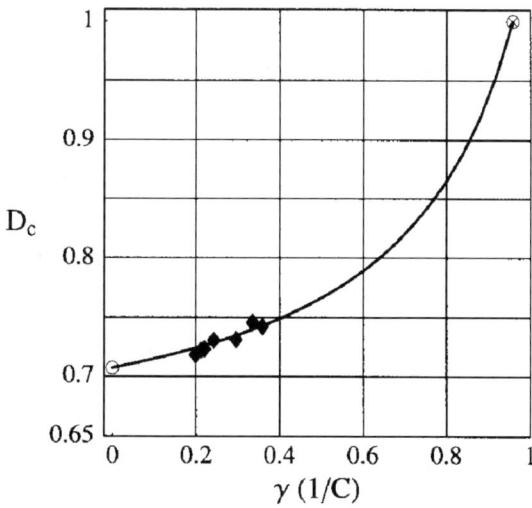

Fig. 21. Comparison of theory and experiment: D_c vs. $\gamma \, [l/\mathrm{mol}]$ curve Solid lines (—): Eq. (102). (○): classical gel points; (⊗): critical dilution. ... Experimental points: by Ross-Murphy (◆: $\kappa = 1$)

Substituting these values into Eq. (102) ($\kappa = 1$), we can now plot D_c as a function of γ (Fig. 21). Agreement of the theory (solid line) with the experimental points (◆) by Ross-Murphy is remarkably good, giving a critical dilution $\gamma_c = 0.96$, greater than that in the adipic acid-pentaerythritol polymer system, which can be ascribed to the lower production of rings in this polymer system (Table 3).

7.1.3
Comparison with Viscoelastic Method

The theory [Eq. (102)] is examined with another experiment with a different polymer mixture. The recent development of the rheological characterization of polymer solutions has enabled experimentalists to identify the gel point [62, 64–68]. Muller and coworkers [67] applied this new method to the gelation of a *polyethylene oxide* ($g = 2$, $\overline{M}_n \cong 1000$) – *Desmodur RF* ($f - g = 3$, $M_n = 465$) mixture [90] diluted by *dioxane* (Fig. 22), and they observed $D_c = 0.71$.

We can now estimate the theoretical gel point in this system: the mean molecular mass of this mixture is calculated to be $\frac{3}{5} \cdot 1000 + \frac{2}{5} \cdot 465 = 786 \, (g/mol)$ for the equimolar case ($\kappa = 1$). Considering the dilution effect ($33 \, w/w \, \%$) by *dioxane*, it follows that $\gamma \approx 2.4 \, l/mol$. The *polyethylene oxide* molecule of interest is composed of $\xi \approx 70$ skeletal bonds on average. Due to this long backbone along with the dense solution, it can be approximated that the structural unit follows Gaussian statistics. With the Flory characteristic constant [6], $C_n = \langle r_n^2 \rangle / n\ell^2 \cong 4$, for *polyethylene oxide* in the Θ regime, one has

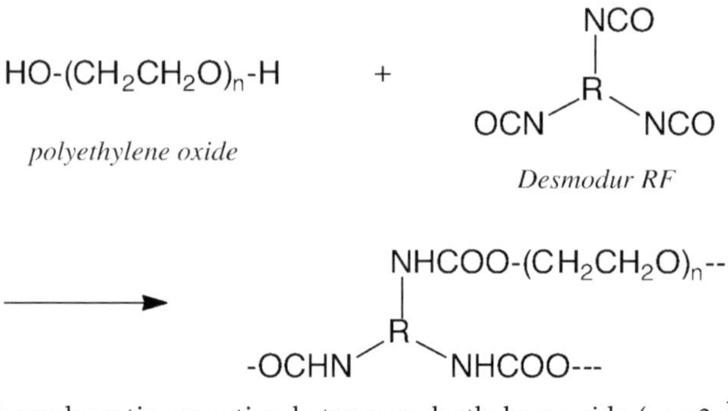

HO-(CH₂CH₂O)ₙ-H   +   Desmodur RF (OCN-R(NCO)-NCO)

polyethylene oxide

⟶   NHCOO-(CH₂CH₂O)ₙ--- / -OCHN-R-NHCOO---

Fig. 22. A polycondensation reaction between polyethylene oxide ($g=2, \overline{M}_n \cong 1000$) and Desmodur RF ($f-g=3$, $M_n = 465$)

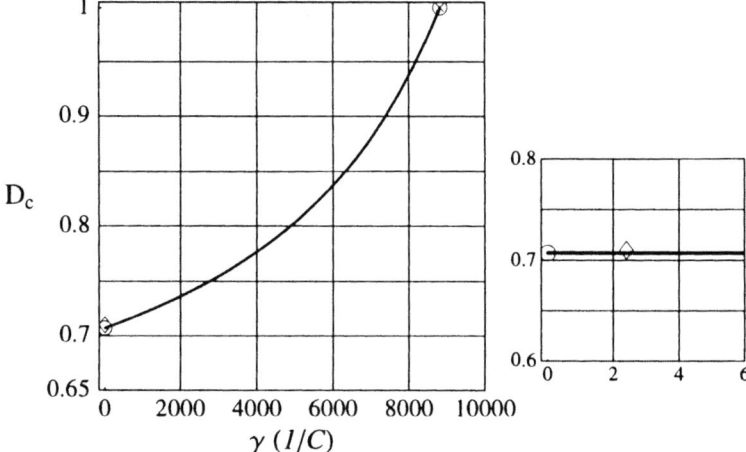

Fig. 23. Comparison of theory and experiment: D_c vs. $\gamma \, [l/\text{mol}]$ curve
The figure shown in the right side box is a magnification of the same plot. Solid lines (—): Eq. (102). (○): classical gel points; (⊗): critical dilution. Experimental points: by Muller and coworkers [67] (◇: κ = 1)

$$\sum_{j=1}^{\infty} \varphi_j / 2j \cong 1.5 \times 10^{-5} \quad \text{and} \quad \sum_{j=1}^{\infty} \varphi_j \cong 6.0 \times 10^{-5}. \tag{117}$$

As these extremely small φ_j values indicate, the production of rings is nearly negligible for this system. Hence, without any further calculation, one recovers the classical value, $D_c = 0.707$, in agreement with the Muller and coworkers' observation $D_c = 0.71$ [67]. For comparison, in Fig. 23 is plotted the theoretical line (solid line) calculated according to Eq. (102), together with the experimental point (◇) by Muller and coworkers.

A characteristic aspect with this branched system is that the critical dilution, γ_c, is located at very high dilution, $\gamma_c \approx 10^4$ (Table 3). From this example, we re-

Table 3. Calculated and observed φ_j and γ_c values

Branched Polymer	Theoretical			Observed		
	$\sum_{x=1}^{\infty}\varphi_x/2x$	$\sum_{x=1}^{\infty}\varphi_x$	Y_c	$\sum_{x=1}^{\infty}\varphi_x/2x$	$\sum_{x=1}^{\infty}\varphi_x$	Y_c
Linear Polyester (Jones and coworkers)				0.066 ~0.10	0.488 ~0.83	
Adipic acid-Pentaerythritol	0.32	4 × 0.32	0.45			$(\approx 0.4)^a$
1,3,5-Tris(carboxymethyl) benzene -Decamethyleneglycol	0.085	8 × 0.085	0.96			
polyethylene oxide-Desmodur RF	1.5 × 10^{-5}	6 × 10^{-5}	≈10^4			

a: Extrapolated to $D_c = 1$.

alize that the critical dilution, γ_c, varies inversely in proportion to the relative cyclization frequency, φ_j, namely, the product, $\Sigma_j^{\infty}\varphi_j \times \gamma_c$ always remains in the order of unity, ≈1. This brings us to the expectation that the experimental determination of γ_c may provide us with useful information about the relative cyclization frequency in branching media.

7.2
Percolation Model

Thanks to a great deal of effort by physicists, the determination of the threshold now amounts to an accuracy higher than 10^{-5}. Although, because of the difference in chemical machinery, one cannot directly test the polymer gel theory by the percolation simulation, one can test the soundness of the underlying mathematical method by way of the examination of the corresponding threshold Eq. (114) derived in a like manner:

$$D_c = \frac{1}{(f-1)p_s}\left\{\frac{1-\sum_j^{\infty}(1-1/j)\mathcal{P}}{1-\sum_j^{\infty}\mathcal{P}}\right\}, \tag{114}$$

Prior to the comparison with simulation experiments, it is necessary to calculate the cyclization probability, $\mathcal{P}$, which is the only unknown parameter with Eq. (114). Note that Eq. (114) was derived based on the 1/d expansion, so that one can apply Gaussian statistics to $\mathcal{P}$. According to the definition, $\mathcal{P}$ has the form:

$$\mathcal{P} = \int_0^{\ell} S_d P(r) dr, \tag{35}$$

where ℓ is the size of a unit cell (bond length). Let n_j be the number of bonds constituting a j-chain, and let a j-chain be comprised of j sites, then one has $n_j = j-1$. Only even number rings can occur for hypercubic lattices. One has, therefore, $n_j = 3, 5, 7, \cdots, 2k+1, \cdots$, for $j = 4, 6, 8, \cdots$. Through a simple vectorial argument, it is easy to show that the average end-to-end distance of j-chains without excluded volume effects has the form[4]:

$$\langle r_j^2 \rangle \doteq \frac{f}{f-2} n_j \ell^2 - \frac{2(f-1)}{(f-2)^2} \ell^2, \quad \text{(hypercubic lattices)}. \tag{118}$$

Let

$$\langle r_j^2 \rangle = v\,\ell^2, \tag{119}$$

---

4 **Mean Square End-to-End Distance of a Chain on Hypercubic Lattices.** Let us consider a random flight chain on hypercubic lattices. Let $\vec{r}_n$ be a vector from a given site to the nth site. Then we have

$$\vec{r}_n = \sum_{i=1}^{n-1} \vec{\ell}_i. \tag{118a}$$

$$\vec{r}_n \cdot \vec{r}_n = (n-1)\ell^2 + 2\sum_{j=2}^{n-1}\sum_{i=1}^{j-1} \vec{\ell}_i \cdot \vec{\ell}_j. \tag{118b}$$

Taking an average, we have

$$\langle \vec{r}_n \cdot \vec{r}_n \rangle = (n-1)\ell^2 + 2\sum_{j=2}^{n-1}\sum_{i=1}^{j-1} \langle \vec{\ell}_i \cdot \vec{\ell}_j \rangle. \tag{118c}$$

Note that

$$\langle \vec{\ell}_i \cdot \vec{\ell}_j \rangle = \beta^{j-i}\ell^2, \tag{118d}$$

for hypercubic lattices, where $\beta = \dfrac{1}{f-1}$. Hence,

$$\langle r_n^2 \rangle = (n-1)\ell^2 + 2\sum_{j=2}^{n-1}\sum_{i=1}^{j-1}\beta^{j-i}\ell^2, \tag{118e}$$

which yields

$$\langle r_n^2 \rangle = \left(\frac{1+\beta}{1-\beta}\right)(n-1)\ell^2 - \frac{2(\beta-\beta^n)}{(1-\beta)^2}\ell^2, \tag{118f}$$

whence we have

$$\langle r_n^2 \rangle = \left(\frac{f}{f-2}\right)(n-1)\ell^2 - \frac{2}{(f-2)^2}\left\{f - 1 - \left(\frac{1}{f-1}\right)^{n-2}\right\}\ell^2, \tag{118g}$$

which leads to the expression of Eq. (118).

and

$$v = \frac{f}{f-2}n_j - 2\frac{(f-1)}{(f-2)^2}, \tag{120}$$

then $\mathcal{P}$ reduces to

$$\mathcal{P} = 1 - \frac{\Gamma\left(\frac{d}{2}, \frac{d}{2v}\right)}{\Gamma\left(\frac{d}{2}\right)}, \tag{35'}$$

where $\Gamma\left(\frac{d}{2}, \frac{d}{2v}\right)$ represents an incomplete Gamma function as defined earlier (see Sect. 4).

7.2.1
Pure Bond Percolation ($p_s = 1$)

Using Eqs. (120) and (35'), one can compute the bond percolation threshold following Eq. (114). In Fig. 24 (a) is plotted the theoretical line (heavy line) together with the observed values ($\diamond$: $p_s = 1$) known by the percolation simulation [79]. The theory agrees remarkably with the simulation points in higher dimensions, but fails in lower dimensions. To show up the discrepancy in low dimensions, the same data are plotted in Fig. 24 (b) in terms of scaled critical points $(D_c - D_{co})/D_{co}$. Comparing the two lines, one finds that the theoretical line abruptly merges with the observed line right at $d = 8$, showing the existence of a critical dimension at $d_c = 8$, above which the foregoing approximations (Assumption I and the high dimension expansion) work well. Eq. (114) is thus exact above 8 dimensions. The result is consistent with the prediction by the Lubensky field theory[30–32]. This is an example of the marginal dimensionality prominent in modern statistical physics.

7.2.2
Bond Percolation ($p_s \neq 1$)

Eq. (114) is a function not only of d, but also of the site fraction, p_s. Hence, one can compare the theory with the general bond percolation ($p_s \neq 1$) [60]. Examples of $d = 3$ to 6 computed in the same manner are plotted in Fig. 25 against the reciprocal of p_s, together with the simulation results by Stauffer ($\diamond$: $d = 3$; $\triangle$: $d = 4$; $\square$: $d = 5$; $\times$: $d = 6$) [58, 106]. The theoretical lines (solid lines for $d = 3$ and 5; broken lines for $d = 4$ and 6) are in good conformity with the observed points by Stauffer. To date, within our knowledge, the corresponding site-bond diagrams in higher dimensions have not been investigated, so that at present we cannot test Eq. (114) in these regimes. However, in light of the critical dimension concept, it is expected that better agreement between the theory and simulation experiments will be observed for $d \geq 8$.

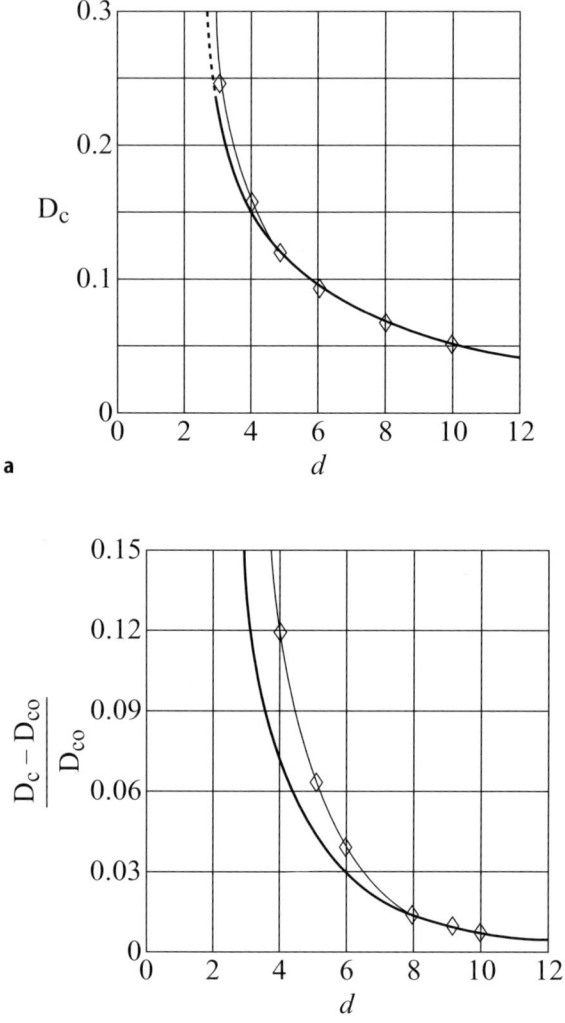

Fig. 24. Comparison of theory and experiment: ($p_s = 1$). (*a*) D_c vs. *d* curve. Theoretical line (heavy line) by Eq. (114), and bond percolation thresholds (◇) cited from ref. [79]. (*b*) Scaled critical point vs. *d* curve. Theoretical line (heavy line): Eq. (114). Bond percolation thresholds: (◇) cited from ref. [79]

7.2.3
Critical Dimension

For some time, the critical dimension of sol clusters on lattices had been thought to be six-dimensions [31, 77–78]. However, when $d_c = 8$ was suggested by Lubensky and Issacsson [32] for the dilution limit of real systems, equivalent to sol

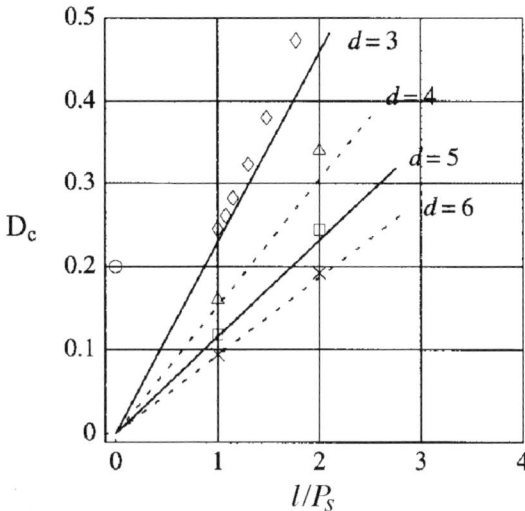

Fig. 25. Comparison of theory and experiment: ($p_s \neq 1$). Theory: Eq. (114) (solid lines for $d=3$ and 5; broken lines for $d=4$ and 6). Simulation points by Stauffer ($\Diamond$: $d=3$; $\triangle$: $d=4$; $\square$: $d=5$; $\times$: $d=6$; refs. [58 and 106]. The open circle ($\bigcirc$) is the corresponding classical threshold, $1/(2d-1)$; i.e., $1/5$ for $d=3$

clusters on lattices, the situation changed to go on in the correct direction. Inspired by the Lubensky theory, Gaunt [33] scrutinized the bond animals up to 10-clusters; applying the numerical values to the asymptotic equality,

$$N_b \sim b^{-\theta} \lambda^b \quad (\text{for } b \to \infty),$$

(N_b is the number of animals with b bonds), and making use of the aforementioned series expansion (Sect. 6), he could show that the exponent θ collapses into the classical value, $\theta = 5/2$, at 8 dimensions, in support of the Lubensky field theory. The Parisi and Sourlas $\varepsilon = 8-d$ expansion [34] gave almost identical results for θ, thus supporting $d_c = 8$ as well. The Isaacsson-Lubensky mean-field theory [30] based on the Flory excluded volume theory and the screening concept is more intelligible, which reads

$$v = \frac{5}{2(d+2)},$$

via the Parisi-Sourlas scaling relation of magnetism [34], giving

$$\theta = \frac{7d-6}{2(d+2)},$$

which reveal more vividly $d_c = 8$; i.e., both the exponents equally collapse into the respective classical values, 1/4, and 5/2, right at 8 dimensions. From these examples, we learn that the ideal behavior is realized above 8 dimensions[5].

7.3
Conclusions

The only unknown parameter of the theoretical Eq. (102) is the relative cyclization frequency, φ_j. For longer chains, this can be computed precisely making use of the Gamma function

$$\varphi_j = \frac{d}{2\pi^{d/2} \ell^d} \left[\Gamma\left(\tfrac{d}{2}\right) - \Gamma\left(\tfrac{d}{2}, \tfrac{d}{2v}\right) \right].$$

For shorter chains, plenty of the observed data are available for extensive polymer species. As a result, theoretical gel points can be estimated within the experimental errors of φ_j. Agreement between the theory and observed values is excellent, in support of the mathematical soundness of Eq. (102). It, however, does not necessarily follow that the present result is the final demonstration of the theory, further examination is still essential.

For the percolation model, agreement of the theory with observations is excellent for $d \geq 8$, showing that the Gaussian approximation works well in high dimensions. However, this is not the case in low dimensions. The agreement at eight-dimensions appears abrupt, suggesting the existence of the marginal dimensionality, $d_c = 8$, in harmony with the Lubensky field theory. The site-bond percolation line shown by Eq. (114) is consistent with Stauffer's simulation experiments which show that the threshold changes linearly in proportion to the reciprocal of the site occupation probability:

$$D_c \propto 1/p_s.$$

---

[5] In a recent work [107], Bunde, Havlin and Porto could generate, using a new growth model, a branched polymer on a lattice, and observed that their branched polymer belongs to the universality class of percolation rather than the lattice animal type. At first sight, their result opposes all existing theories and observations [30–34]. It is important to note that their branched polymer is of an addition type, while almost all discussions about the critical behavior have been done for a condensation type polymer that grows via the head-to-tail reaction

$$L_j = \sum_{i=1}^{j-1} L_i L_{j-i}$$

whose profile is of the Kurata type shown in Fig. 2, where L_j represents a j-cluster. If the Bunde, Havlin and Porto result is final, their polymer will belong to another universality class different from the Kurata type [13, 108–109].

In light of the concept of the marginal dimensionality, it is expected that this hyperbolic relationship holds more accurately above eight dimensions.

8
Expectations of Future Theories

Through the examinations in Sect. 7 it was found that the gel point Eq. (102) of real systems is in excellent agreement with all the experiments taken up in this article (Figs. 19, 21 and 23), while the corresponding Eq. (114) for the percolation model was exact in high dimensions, but failed in low dimensions; substantial deviations were found for $d<6$ (Fig. 24). Those gel point equations are founded on the criteria: (1) the assumption of random distribution of cyclic bonds, and (2) the series expansion of $[\Gamma]$, together with the premises of Eq. (96) for real systems and of Eq. (110) for the percolation model. At present, as to real systems no inconsistency has been found that disturbs these criteria, suggesting that the Eqs. (101)–(103) have sound mathematical basis. The situation is different, on the other hand, for the percolation model, since the monomer molecules are fixed on lattices (Sect. 5). Indeed the result in Fig. 24 shows that the critera do not work in low dimensions. The observed discrepancy in low dimensions may have two origins:

First, ring formation on lattices can take place on even clusters alone larger than 4, say, for hypercubic lattices. Thus, cyclization is not a random event, but depends on cluster size. One can check this possibility by enumerating the number of rings on lattices. Now pay attention to the Eq. (108) which holds exactly only if the random distribution assumption holds. The enumeration in the two dimensional bond percolation showed that rings are considerably fewer than expected from Eq. (108); that is, $D(inter) > 1/(f-1)$. Thus, the random distribution assumption is violated for low dimensional bond percolations. This may be an essential part of the discrepancy observed in low dimensions, which, however, cannot explain the sudden confluence at $d=8$, since $D(inter)$ is in nature independent of d in the percolation model (Sect. 5) [84].

Second, the failure of the $D(ring)$ term should arise from the high dimension expansion based on the premise of Eq. (110). The result of Fig. 24 shows that the expansion works well for $d \geq 8$. It has been well established that lattice branched clusters have the marginal dimensionality, $d_c = 8$, in the sol phase (bond animals) above which the ideal behavior applies [33]. Now, the critical point shift results from ring formation which is a phenomenon in the sol phase up to the gel point, and the cyclization frequency is influenced by the excluded volume effects. The gel point, therefore, must shift in response to the behavior of the sol clusters. This leads us to a conjecture that a mathematical singularity may arise at $d_c = 8$ on the D_c vs. d curve, in parallel with the phase transition from the excluded volume clusters to the ideal ones. If this is the case, it follows that the high dimension expansion must fail below eight dimensions. To date, there is no experimental evidence that shows the existence of the discontinuity on the D_c vs. d curve, but it is likely that D_c is not a monotonous function of d; the result of Fig.

24 suggests this possibility. At first sight, the idea seems odd that there may exist a critical dimension of critical points above which the classical value is not recovered as opposed to the case of (critical) exponents [20]. However, all evidence seems to suggest the existence of the singularity.

Whether the singularity occurs or not, the answer to the question, *why is the critical point located there*, should reside within the framework of the basic Eqs. (91) and (105). When these equalities are solved exactly, then we will be able to take one step forward in this exciting field of polymer science.

The most characteristic aspect of the critical point problem is that the three phenomena, cyclization, excluded volume effects, and dimension, intimately interacting with each other, spontaneously appear at the critical point. At the beginning, it was thought that cyclization would make little contribution to such an important question that has remained unsolved for so long in physical science. The author's early conjecture was wrong. As we have seen in the text, cyclization plays a central role in the location of the critical point. For the percolation model, dimension is almost equivalent to cyclization (Sects. 4 and 5); even excluded volume effects seem to manifest themselves as an element of cyclization (Sects. 6 and 7), while dimensionality is in close conjunction with excluded volume effects (Sect. 7). In real gelations, the three effects are deeply connected with one another.

According to the excluded volume theory [30] combined with the screening concept [19], it is inferred that branched molecules in concentrated sol phase should have a critical dimension, $d_c < 8$, lower than that for the percolation clusters ($d_c = 8$). For the percolation model, our world ($d = 3$) is farther away from the critical dimension ($d_c = 8$) than is the case of real systems ($d_c < 8$). This is the reason of the remarkable agreement with the observations in real systems (Figs. 19, 21 and 23), and the large deviations from the low dimensional percolation simulations (Figs. 24 and 25). To accomplish better agreement with observations, one must seek better approximations for the basic equalities (91) and (105). At present, however, there does not seem to be another good idea for this. The author hopes that the next generation will realize this with deeper insights and sophisticated mathematical approaches.

Through the present review, one impressive aspect is that the gel science (perhaps all sciences) looks like a branching process itself: Initially, many theoretical and experimental researches rise independently. Then those interact with each other, collide, and are activated, and some of them, taking in different findings from other fields, grow and branch off to larger clusters forming linkages with others. Those finally merge into a bigger molecule, in which all sorts of intellectual information are being condensed. Macroscopically, this can be observed as homogeneous, while microscopically it is filled with wild inhomogeneity and fluctuations. This is just a network molecule itself.

Acknowledgements. The author greatly thanks Professor Dusek, the editor, for his kind invitation for this review article, Doctor Spouge (NIH) for his valuable information about gel theories, and Doctor Kawamata (Keio University) for useful suggestions about steric effects of ring molecules. The author is grateful to John Wiley & Sons, Inc. for the permission to reproduce Figs. 13 and 14 from the Journal of Applied Polymer Science, 27, 571 and 572, 1982.

References

1. Carothers WH (1931) Chem Rev **8**: 353
2. Flory PJ (1941) J Am Chem Soc **63**: 3083
3. Lee S (1939) Kasen Kouen Yousisyu (Lectures in Chemical Fibers) **4**: 51
4. Herrmann WO, Haehnel W (1927) Ber dtsch chem Ges **60**: 1658
5. Staudinger H, Frey K, Starck W (1927) Ber dtsch chem Ges **60**: 1782
6. Flory PJ (1953) Principles of Polymer Chemistry. Cornell University Press, Ithaca and London; Flory PJ (1969) Statistical Mechanics of Chain Molecules, John Wiley & Sons, Inc, New York
7. Good IJ (1948) Proc Camb Phil Soc **45**: 360; (1955) ibid **51**: 240; (1960) ibid **56**: 367; (1963) Proc Roy Soc **A292**: 54
8. Dobson GR, Gordon M (1964) J Chem Phys **41** No. 8: 2389
9. Dusek K, Gordon M, Ross-Murphy SB (1978) Macromolecules **11**: 236
10. Dusek K (1984) Macromolecules **17**: 716
11. Kajiwara K (1971) Polymer **12**: 57
12. Suematsu K (1992) J Phys Soc Japan **61**: 1539
13. Kurata M, Fukastsu M (1964) J Chem Phys **41**: 2934
14. Burchard W (1983) Advances in Polymer Sciences **48**: 1
15. Reif R (1985) Fundamentals of Statistical and Thermal Physics, McGraw-Hill, New York
16. Pauling L (1960) The Nature of the Chemical Bonds. Cornell University Press, Ithaca, New York
17. Atkins PW (1994) Physical Chemistry. Oxford University Press, Oxford
18. Fisher M (1966) J Chem Phys **44**: 616
19. de Gennes PG (1979) Scaling Concepts in Polymer Physics, Cornell University Press, Ithaca
20. Zallen R (1983) The Physics of Amorphous Solids. Wiley, New York
21. Macdonald D, Hunter DL, Kelley K, Jan N (1992) J Phys **A 25**: 1429
22. Fixman M (1992) J Phys **A 25**: 1429
23. Edwards SF (1965) Proc Phys Soc **85**: 613
24. Domb C, Gillis J, Wilmers G (1965) Proc Phys Soc **85**: 625
25. Domb C (1969) Adv Chem Phys **15**: 229
26. Kirkpatric S (1976) Phys Rev Letters **36**: 69
27. Debye PJW, Hückel E (1923) Physik Z **24**: 185, 305
28. Debye PJW (1924) Physik Z **25**: 97
29. Chu B (1969) Molecular Forces. Based on the Baker Lectures of Peter J. W. Debye, John Wiley & Sons, New York
30. Isaacson J, Lubensky TC (1980) J Physique Letters **41**: L-469
31. Lubensky TC (1978) Physical Review Letters **41**: 829
32. Lubensky TC, Isaacson J (1979) Physical Review A **20**: 2130
33. Gaunt DS (1980) J Phys A Math. Gen **13**: L-97
34. Parisi G, Sourlas N (1981) Phys Rev Letters **46** No14: 871
35. Toulouse G (1974) Nuovo Cimento **23**: 234
36. Ziman T (1979) Ordering in Strongly Fluctuating Condensed Matter Systems, Riste T (ed), Plenum Press, London
37. des Cloizeaux J (1980) J Physique **41**: 223
38. Bishop M, Saltiel CJ (1988) J Chem Phys **89**: 1159
39. Bishop M, Clarke JHR (1989) J Chem Phys **91**: 3721, 6345
40. Pereira GG (1995) Physica A **219**: 290

41. Valleau JP (1996) J Chem Phys **104**: 3071
42. Suematsu K (1993) J Chem Soc Faraday Trans **89**: 4181
43. Macosko CW, Miller DR (1976) Macromolecules **9**: 199
44. Miller DR, Macosko CW (1976) Macromolecules **9**: 206
45. Landin DT, Macosko CW (1988) Macromolecules **21**: 646
46. Dotson NA, Galvan R, Macosko CW (1988) Macromolecules **21**: 2560
47. Suematsu K (1996) J Chem Soc Faraday Trans **92**: 2417
48. Spouge JL (1983) J Stat Phys **31**: 363; (1986) **43**: 143; (1983) Macromolecules **16**: 121
49. Suematsu K, Okamoto T (1992) J Stat Phys **66**: 661
50. Suematsu K, Kohno M (1995) J Theor Biol **175**: 317
51. Ziman J (1979) Models of disorder, Cambridge University Press, New York
52. Klein DJ, Seitz WA (1983) Chemical Applications of Topology and Graph Theory, A Collection of Papers from a Symposium Held at the University of Georgia, Athens, Georgia, USA, 18-22 April, R. B. King (ed), Studies in Physical and Theoretical Chemistry **28**: 430 Elsevier Science Publishers, Amsterdam.
53. Gaylord RJ, Wellin PR (1997) Computer Simulations with MATHEMATICA, Explorations in Complex Physical and Biological Systems. TELOS, The Electronic Library of Science, Santa Clara, California
54. Herrmann HJ (1986) Physics Reports **136**: 153
55. Martin JL, Sykes MF, Hioe FT (1967) J Chem Phys **46**: 3478
56. Domb C, Gillis J, Wilmers G (1965) Proc Phys Soc **85**: 625
57. Suematsu K, Kohno M (1998) J Stat Phys **93**: 293
58. Stauffer D (1982) Adv Polymer Sci **44**: 103; (1985) Introduction to Percolation Theory. Taylor & Francis, London
59. Brinker CJ, Scherer GW (1990) Sol-Gel Science, The Physics and Chemistry of Sol-Gel Processing. Academic Press, Harcourt Brace Javanovich, Boston
60. Agrawal P, Redner S, Reynolds PJ, Stanley HE (1979) J Phys A Math. Gen **12**: 2073
61. Nakanishi H, Reynolds PJ (1979) Phys Letters **71A**: 252
62. Tung CM, Dynes PJ (1982) J Appl Polym Sci **27**: 569
63. ASTM D2471-71 (1971) Gel Time and Peak Exothermic Temperature of Reacting Thermosetting Reins
64. Chambon F, Winter HH (1985) Polym Bull **13**: 499; (1986) J Rheol **31**: 683
65. Winter HH (1997) Adv Polym Sci **134**: 165
66. Winter HH, Chambon F (1986) J Rheol **30**: 367
67. Muller R, Gerard E, Dugand P, Rempp P, Gnanou Y (1986) J Rheol **30**: 367
68. Lin HL, Yu TL, Cheng CH (1999) Macromolecules **32**: 690
69. Sykes MF, Essam JW (1963) Phys Rev Lett **10**: 3; (1964) J Math Phys **5**: 1117
70. Shante VKS, Kirkpatrick S (1971) Adv Phys **20**: 325
71. Domb C, Sykes MF (1961) Phys Rev **122**: 77
72. Domb C (1983) Ann Israel Phys Soc, Percolation Structures and Processes, 5: 17 G. Deutscher, Zallen R, Adler J (ed)
73. Sykes MF, Essam JW (1964) Phys Rev **133**: A310
74. Essam JW (1972) Phase Transitions and Critical Phenomena. Domb C, Green M (eds), Vol. 2: 192, Academic Press, New York
75. Moriguchi S, Udagawa K, Ichimatsu S (1987) Mathematical Formulae. Vol. 2: 104 Printed by Iwanami, Tokyo, Japan
76. Sykes MF, Martin JL, Essam JW (1973) J Phys A Math Nucl Gen **6**: 1306
77. Gaunt DS, Sykes MF, Ruskin H (1976) J Phys A Math Gen **9**: 1899
78. Gaunt DS, Ruskin H (1978) J Phys A Math Gen **11**: 1369
79. Galam S, Mauger A (1994) Physica **A205**: 502; (1996) Phys Rev **E53**: 2177; (1997) Phys Rev **E55**: 1230; (1998) Eur Phys J **B1**: 255
80. van der Marck SC (1997) Phys Rev **E55**: 6593; (1997) Phys Rev **E55**: 1514; (1997) Phys Rev **E55**: 1228; (1998) J Phys A Math Gen **31**: 3449; (1998) Int J Mod Phys **9**: 529
81. Wile LL (1945) Ph.D. dissertation, Columbia University, New York

82. Spouge JL (1985) J Phys A Math Gen **18**: 3063; (1984) Adv Appl Prob **16**: 275
83. Suematsu K (1998) Eur Phys J **B6**: 93; (1998) Theoretical View of Physical Gels, Kobunshi Ronbunshu **55**: No12, 723
84. Suematsu K (2000) Phys Rev **E62**: 3944
85. Dusek K, Vojta V (1977) Br Polym J **9**: 164
86. Dusek K, Ilavsky M (1975) J Polym Sci Symposium No53: 57, 75
87. Dusek K (1979) Makromol Chem Suppl **2**: 35; (1982) Development in Polymerization 3, Haward RN (ed), Elsevier Applied Science Publishers Ltd London p-143
88. Somvarsky J, Dusek K (1994) Polym Bull **33**: 369, 377
89. Dotson NA, Macosko CW, Tirrell M (1992) Synthesis, Characterization and Theory of Polymeric Networks and Gels. Aharoni A (ed), Plenum Press, New York
90. Hild G (1998) Prog Polym Sci **23**: 1019
91. Clark AH (1995) Faraday Discuss **101**: 77
92. Matejka L, Dusek K (1980) Polym Bull **3**: 489
93. Broadbent SR, Hammersley JM (1957) Proc Camb Phil Sci **53**: 629
94. Frisch HL, Hammersley JM (1963) J Soc Indust Appl Math **11**: 894
95. Stauffer D (1976) J Chem Soc Faraday Trans II**72**: 1354
96. de Gennes PG (1976) J Physique (Letters) **37**: L-1
97. Vyssotsky VA, Gordon SB, Frisch HL, Hammersley JM (1961) Phys Rev **123**: 1566
98. Gordon M, Scantlebury GR (1967) J Chem Soc (**B**): 1
99. Ross-Murphy SB (1975) J Polymer Sci Symp **53**: 11
100. Jones FR, Scales LE, Semlyen JA (1974) Polymer **15**: 738
101. Cooper DR, Semlyen JA (1973) Polymer **14**: 185
102. Semlyen JA (1982) Copolymerization In: Adv in Polymerization 3, Haward RN (ed), Elsevier Applied Science Publishers Ltd London; (1986) Cyclic Polymers, Elsevier, London; (1996) Large Ring Molecules. Wiley, New York 103. Semlyen JA
103. Ingold CK (1969) Structure and Mechanism in Organic Chemistry. Cornell University Press, Ithaca
104. Eliel EL, Allinger NL, Angyal ST, Morrison GA (1965) Conformational Analysis, Chap 4, Interscience Publishers, New York
105. Ruzicka L, Stoll M, Schinz H (1926) Helv Chim Acta **9**: 249
106. Gropengiesser U, Stauffer D (1994) Physica **A210**: 320
107. Bunde A, Havlin S, Porto M (1995) Phys Rev Letters **74**: 2714
108. Herrmann HJ, Landau DP, and Stauffer D (1982) Phys Rev Letters **49**: 412
109. Schmidt M, Burchard W (1981) Macromolecules **14**: 370

Appendix

According to the Flory excluded volume theory and the de Gennes screening effect, the exponent of the molecular dimension is given by the Isaacson-Lubensky formula:

$$v = \frac{2(1+v_0)-\varrho}{d+2}, \qquad (27)$$

where v_0 is the exponent for ideal molecules. Typical cases for a chain and a branched molecule are shown in Table 4. Consider a $d = 3$ case of a chain molecule, and Eq. (27) yields 3/5 for the dilution limit ($C = 0$) and 2/5 for the melt. The former value is the well-established Flory exponent, while the latter is smaller than the observed value, 1/2, the Gaussian exponent. Thus it seems natural to infer that, somewhere between $0 < C < C_{melt}$, there exists a critical concentra-

Table 4.

	ν	
	$C = 0$	$C = C_{melt}$
chain	$\dfrac{3}{d+2}$	$\dfrac{2}{d+2}$
branched	$\dfrac{5}{2(d+2)}$	$\dfrac{3}{2(d+2)}$

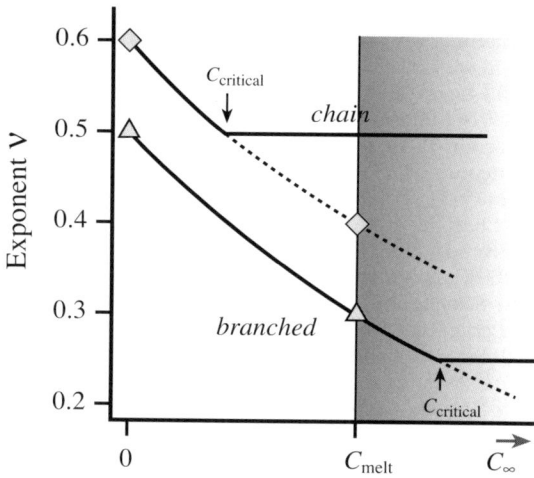

Fig. 26. Concentration dependence of the exponent ν. Critical concentration concept

tion C_{crit} where sudden change from a excluded volume chain to the ideal chain occurs; i.e., it is not unlikely that the change from a excluded volume molecule to the ideal one is a transition like the dimension-dependence of ν.

For a branched molecule, ν takes the value $3/10$ for C_{melt} greater than that for the ideal case $\nu_0 = 1/4$. In this case, it is expected that the same transition occurs at a certain hypothetical regime of $C_{crit} > C_{melt}$ (Fig. 26). The schematic representation is given in Fig. 26.

Editor: K. Dušek
Received: November 2000

Author Index Volumes 101–156

Author Index Volumes 1–100 see Volume 100

de, Abajo, J. and *de la Campa, J.G.*: Processable Aromatic Polyimides. Vol. 140, pp. 23-60.
Adolf, D. B. see *Ediger, M. D.*: Vol. 116, pp. 73-110.
Aharoni, S. M. and *Edwards, S. F.*: Rigid Polymer Networks. Vol. 118, pp. 1-231.
Améduri, B., Boutevin, B. and *Gramain, P.*: Synthesis of Block Copolymers by Radical Polymerization and Telomerization. Vol. 127, pp. 87-142.
Améduri, B. and *Boutevin, B.*: Synthesis and Properties of Fluorinated Telechelic Monodispersed Compounds. Vol. 102, pp. 133-170.
Amselem, S. see *Domb, A. J.*: Vol. 107, pp. 93-142.
Andrady, A. L.: Wavelenght Sensitivity in Polymer Photodegradation. Vol. 128, pp. 47-94.
Andreis, M. and *Koenig, J. L.*: Application of Nitrogen-15 NMR to Polymers. Vol. 124, pp. 191-238.
Angiolini, L. see *Carlini, C.*: Vol. 123, pp. 127-214.
Anseth, K. S., Newman, S. M. and *Bowman, C. N.*: Polymeric Dental Composites: Properties and Reaction Behavior of Multimethacrylate Dental Restorations. Vol. 122, pp. 177-218.
Antonietti, M. see *Cölfen, H.*: Vol. 150, pp. 67-187.
Armitage, B. A. see *O'Brien, D. F.*: Vol. 126, pp. 53-58.
Arndt, M. see *Kaminski, W.*: Vol. 127, pp. 143-187.
Arnold Jr., F. E. and *Arnold, F. E.*: Rigid-Rod Polymers and Molecular Composites. Vol. 117, pp. 257-296.
Arshady, R.: Polymer Synthesis via Activated Esters: A New Dimension of Creativity in Macromolecular Chemistry. Vol. 111, pp. 1-42.

Bahar, I., Erman, B. and *Monnerie, L.*: Effect of Molecular Structure on Local Chain Dynamics: Analytical Approaches and Computational Methods. Vol. 116, pp. 145-206.
Ballauff, M. see *Dingenouts, N.*: Vol. 144, pp. 1-48.
Baltá-Calleja, F. J., González Arche, A., Ezquerra, T. A., Santa Cruz, C., Batallón, F., Frick, B. and *López Cabarcos, E.*: Structure and Properties of Ferroelectric Copolymers of Poly(vinylidene) Fluoride. Vol. 108, pp. 1-48.
Barnes, M. D. see *Otaigbe, J.U.*: Vol. 154, pp. 1-86.
Barshtein, G. R. and *Sabsai, O. Y.*: Compositions with Mineralorganic Fillers. Vol. 101, pp.1-28.
Baschnagel, J., Binder, K., Doruker, P., Gusev, A. A., Hahn, O., Kremer, K., Mattice, W. L., Müller-Plathe, F., Murat, M., Paul, W., Santos, S., Sutter, U. W., Tries, V.: Bridging the Gap Between Atomistic and Coarse-Grained Models of Polymers: Status and Perspectives. Vol. 152, pp. 41-156.
Batallán, F. see *Baltá-Calleja, F. J.*: Vol. 108, pp. 1-48.
Batog, A. E., Pet'ko, I. P., Penczek, P.: Aliphatic-Cycloaliphatic Epoxy Compounds and Polymers. Vol. 144, pp. 49-114.
Barton, J. see *Hunkeler, D.*: Vol. 112, pp. 115-134.
Bell, C. L. and *Peppas, N. A.*: Biomedical Membranes from Hydrogels and Interpolymer Complexes. Vol. 122, pp. 125-176.
Bellon-Maurel, A. see *Calmon-Decriaud, A.*: Vol. 135, pp. 207-226.

Bennett, D. E. see *O'Brien, D. F.:* Vol. 126, pp. 53-84.
Berry, G.C.: Static and Dynamic Light Scattering on Moderately Concentraded Solutions: Isotropic Solutions of Flexible and Rodlike Chains and Nematic Solutions of Rodlike Chains. Vol. 114, pp. 233-290.
Bershtein, V. A. and *Ryzhov, V. A.:* Far Infrared Spectroscopy of Polymers. Vol. 114, pp. 43-122.
Bigg, D. M.: Thermal Conductivity of Heterophase Polymer Compositions. Vol. 119, pp. 1-30.
Binder, K.: Phase Transitions in Polymer Blends and Block Copolymer Melts: Some Recent Developments. Vol. 112, pp. 115-134.
Binder, K.: Phase Transitions of Polymer Blends and Block Copolymer Melts in Thin Films. Vol. 138, pp. 1-90.
Binder, K. see *Baschnagel, J.:* Vol. 152, pp. 41-156.
Bird, R. B. see *Curtiss, C. F.:* Vol. 125, pp. 1-102.
Biswas, M. and *Mukherjee, A.:* Synthesis and Evaluation of Metal-Containing Polymers. Vol. 115, pp. 89-124.
Biswas, M. and *Sinha Ray, S.:* Recent Progress in Synthesis and Evaluation of Polymer-Montmorillonite Nanocomposites. Vol. 155, pp. 167-221.
Bolze, J. see *Dingenouts, N.:* Vol. 144, pp. 1-48.
Boutevin, B. and *Robin, J. J.:* Synthesis and Properties of Fluorinated Diols. Vol. 102. pp. 105-132.
Boutevin, B. see *Amédouri, B.:* Vol. 102, pp. 133-170.
Boutevin, B. see *Améduri, B.:* Vol. 127, pp. 87-142.
Bowman, C. N. see *Anseth, K. S.:* Vol. 122, pp. 177-218.
Boyd, R. H.: Prediction of Polymer Crystal Structures and Properties. Vol. 116, pp. 1-26.
Briber, R. M. see *Hedrick, J. L.:* Vol. 141, pp. 1-44.
Bronnikov, S. V., Vettegren, V. I. and *Frenkel, S. Y.:* Kinetics of Deformation and Relaxation in Highly Oriented Polymers. Vol. 125, pp. 103-146.
Brown, H. R. see *Creton, C.:* Vol. 156, pp. 53-135.
Bruza, K. J. see *Kirchhoff, R. A.:* Vol. 117, pp. 1-66.
Budkowski, A.: Interfacial Phenomena in Thin Polymer Films: Phase Coexistence and Segregation. Vol. 148, pp. 1-112.
Burban, J. H. see *Cussler, E. L.:* Vol. 110, pp. 67-80.
Burchard, W.: Solution Properties of Branched Macromolecules. Vol. 143, pp. 113-194.

Calmon-Decriaud, A. Bellon-Maurel, V., Silvestre, F.: Standard Methods for Testing the Aerobic Biodegradation of Polymeric Materials. Vol 135, pp. 207-226.
Cameron, N. R. and *Sherrington, D. C.:* High Internal Phase Emulsions (HIPEs)-Structure, Properties and Use in Polymer Preparation. Vol. 126, pp. 163-214.
de la Campa, J. G. see *de Abajo, , J.:* Vol. 140, pp. 23-60.
Candau, F. see *Hunkeler, D.:* Vol. 112, pp. 115-134.
Canelas, D. A. and *DeSimone, J. M.:* Polymerizations in Liquid and Supercritical Carbon Dioxide. Vol. 133, pp. 103-140.
Capek, I.: Kinetics of the Free-Radical Emulsion Polymerization of Vinyl Chloride. Vol. 120, pp. 135-206.
Capek, I.: Radical Polymerization of Polyoxyethylene Macromonomers in Disperse Systems. Vol. 145, pp. 1-56.
Capek, I.: Radical Polymerization of Polyoxyethylene Macromonomers in Disperse Systems. Vol. 146, pp. 1-56.
Capek, I. and *Chern, C.-S.:* Radical Polymerization in Direct Mini-Emulsion Systems. Vol. 155, pp. 101-166.
Carlini, C. and *Angiolini, L.:* Polymers as Free Radical Photoinitiators. Vol. 123, pp. 127-214.
Carter, K. R. see *Hedrick, J. L.:* Vol. 141, pp. 1-44.
Casas-Vazquez, J. see *Jou, D.:* Vol. 120, pp. 207-266.
Chandrasekhar, V.: Polymer Solid Electrolytes: Synthesis and Structure. Vol 135, pp. 139-206
Chang, J.Y. see *Han, M. J.:* Vol. 153, pp. 1-36.

Charleux, B., Faust R.: Synthesis of Branched Polymers by Cationic Polymerization. Vol. 142, pp. 1-70.
Chen, P. see Jaffe, M.: Vol. 117, pp. 297-328.
Chern, C.-S. see Capek, I.: Vol. 155, pp. 101-166.
Choe, E.-W. see Jaffe, M.: Vol. 117, pp. 297-328.
Chow, T. S.: Glassy State Relaxation and Deformation in Polymers. Vol. 103, pp. 149-190.
Chung, T.-S. see Jaffe, M.: Vol. 117, pp. 297-328.
Cölfen, H. and *Antonietti, M.:* Field-Flow Fractionation Techniques for Polymer and Colloid Analysis. Vol. 150, pp. 67-187.
Comanita, B. see Roovers, J.: Vol. 142, pp. 179-228.
Connell, J. W. see Hergenrother, P. M.: Vol. 117, pp. 67-110.
Creton, C., Kramer, E. J., Brown, H. R., Hui, C.-Y.: Adhesion and Fracture of Interfaces Between Immiscible Polymers: From the Molecular to the Continuum Scale. Vol. 156, pp. 53-135.
Criado-Sancho, M. see Jou, D.: Vol. 120, pp. 207-266.
Curro, J.G. see Schweizer, K.S.: Vol. 116, pp. 319-378.
Curtiss, C. F. and *Bird, R. B.:* Statistical Mechanics of Transport Phenomena: Polymeric Liquid Mixtures. Vol. 125, pp. 1-102.
Cussler, E. L., Wang, K. L. and *Burban, J. H.:* Hydrogels as Separation Agents. Vol. 110, pp. 67-80.

DeSimone, J. M. see Canelas D. A.: Vol. 133, pp. 103-140.
DiMari, S. see Prokop, A.: Vol. 136, pp. 1-52.
Dimonie, M. V. see Hunkeler, D.: Vol. 112, pp. 115-134.
Dingenouts, N., Bolze, J., Pötschke, D., Ballauf, M.: Analysis of Polymer Latexes by Small-Angle X-Ray Scattering. Vol. 144, pp. 1-48.
Dodd, L. R. and *Theodorou, D. N.:* Atomistic Monte Carlo Simulation and Continuum Mean Field Theory of the Structure and Equation of State Properties of Alkane and Polymer Melts. Vol. 116, pp. 249-282.
Doelker, E.: Cellulose Derivatives. Vol. 107, pp. 199-266.
Dolden, J. G.: Calculation of a Mesogenic Index with Emphasis Upon LC-Polyimides. Vol. 141, pp. 189-245.
Domb, A. J., Amselem, S., Shah, J. and *Maniar, M.:* Polyanhydrides: Synthesis and Characterization. Vol.107, pp. 93-142.
Doruker, P. see Baschnagel, J.: Vol. 152, pp. 41-156.
Dubois, P. see Mecerreyes, D.: Vol. 147, pp. 1-60.
Dubrovskii, S. A. see Kazanskii, K. S.: Vol. 104, pp. 97-134.
Dunkin, I. R. see Steinke, J.: Vol. 123, pp. 81-126.
Dunson, D. L. see McGrath, J. E.: Vol. 140, pp. 61-106.

Eastmond, G. C.: Poly(ε-caprolactone) Blends. Vol.149, pp. 59-223.
Economy, J. and *Goranov, K.:* Thermotropic Liquid Crystalline Polymers for High Performance Applications. Vol. 117, pp. 221-256.
Ediger, M. D. and *Adolf, D. B.:* Brownian Dynamics Simulations of Local Polymer Dynamics. Vol. 116, pp. 73-110.
Edwards, S. F. see Aharoni, S. M.: Vol. 118, pp. 1-231.
Endo, T. see Yagci, Y.: Vol. 127, pp. 59-86.
Engelhardt, H. and *Grosche, O.:* Capillary Electrophoresis in Polymer Analysis. Vol. 150, pp. 189-217.
Erman, B. see Bahar, I.: Vol. 116, pp. 145-206.
Ewen, B, Richter, D.: Neutron Spin Echo Investigations on the Segmental Dynamics of Polymers in Melts, Networks and Solutions. Vol. 134, pp. 1-130.
Ezquerra, T. A. see Baltá-Calleja, F. J.: Vol. 108, pp. 1-48.

Faust, R. see Charleux, B: Vol. 142, pp. 1-70.
Fekete, E see Pukánszky, B: Vol. 139, pp. 109-154.
Fendler, J.H.: Membrane-Mimetic Approach to Advanced Materials. Vol. 113, pp. 1-209.

Fetters, L. J. see *Xu, Z.*: Vol. 120, pp. 1-50.
Förster, S. and *Schmidt, M.*: Polyelectrolytes in Solution. Vol. 120, pp. 51-134.
Freire, J. J.: Conformational Properties of Branched Polymers: Theory and Simulations. Vol. 143, pp. 35-112.
Frenkel, S. Y. see *Bronnikov, S. V.*: Vol. 125, pp. 103-146.
Frick, B. see *Baltá-Calleja, F. J.*: Vol. 108, pp. 1-48.
Fridman, M. L.: see *Terent´eva, J. P.*: Vol. 101, pp. 29-64.
Fukui, K. see *Otaigbe, J. U.*: Vol. 154, pp. 1-86.
Funke, W.: Microgels-Intramolecularly Crosslinked Macromolecules with a Globular Structure. Vol. 136, pp. 137-232.

Galina, H.: Mean-Field Kinetic Modeling of Polymerization: The Smoluchowski Coagulation Equation. Vol. 137, pp. 135-172.
Ganesh, K. see *Kishore, K.*: Vol. 121, pp. 81-122.
Gaw, K. O. and Kakimoto, M.: Polyimide-Epoxy Composites. Vol. 140, pp. 107-136.
Geckeler, K. E. see *Rivas, B.*: Vol. 102, pp. 171-188.
Geckeler, K. E.: Soluble Polymer Supports for Liquid-Phase Synthesis. Vol. 121, pp. 31-80.
Gehrke, S. H.: Synthesis, Equilibrium Swelling, Kinetics Permeability and Applications of Environmentally Responsive Gels. Vol. 110, pp. 81-144.
de Gennes, P.-G.: Flexible Polymers in Nanopores. Vol. 138, pp. 91-106.
Giannelis, E.P., Krishnamoorti, R., Manias, E.: Polymer-Silicate Nanocomposites: Model Systems for Confined Polymers and Polymer Brushes. Vol. 138, pp. 107-148.
Godovsky, D. Y.: Device Applications of Polymer-Nanocomposites. Vol. 153, pp. 163-205.
Godovsky, D. Y.: Electron Behavior and Magnetic Properties Polymer-Nanocomposites. Vol. 119, pp. 79-122.
González Arche, A. see *Baltá-Calleja, F. J.*: Vol. 108, pp. 1-48.
Goranov, K. see *Economy, J.*: Vol. 117, pp. 221-256.
Gramain, P. see *Améduri, B.*: Vol. 127, pp. 87-142.
Grest, G.S.: Normal and Shear Forces Between Polymer Brushes. Vol. 138, pp. 149-184.
Grigorescu, G, Kulicke, W.-M.: Prediction of Viscoelastic Properties and Shear Stability of Polymers in Solution. Vol. 152, p. 1-40.
Grosberg, A. and *Nechaev, S.*: Polymer Topology. Vol. 106, pp. 1-30.
Grosche, O. see *Engelhardt, H.*: Vol. 150, pp. 189-217.
Grubbs, R., Risse, W. and *Novac, B.*: The Development of Well-defined Catalysts for Ring-Opening Olefin Metathesis. Vol. 102, pp. 47-72.
van Gunsteren, W. F. see *Gusev, A. A.*: Vol. 116, pp. 207-248.
Gusev, A. A., Müller-Plathe, F., van Gunsteren, W. F. and *Suter, U. W.*: Dynamics of Small Molecules in Bulk Polymers. Vol. 116, pp. 207-248.
Gusev, A. A. see *Baschnagel, J.*: Vol. 152, pp. 41-156.
Guillot, J. see *Hunkeler, D.*: Vol. 112, pp. 115-134.
Guyot, A. and *Tauer, K.*: Reactive Surfactants in Emulsion Polymerization. Vol. 111, pp. 43-66.

Hadjichristidis, N., Pispas, S., Pitsikalis, M., Iatrou, H., Vlahos, C.: Asymmetric Star Polymers Synthesis and Properties. Vol. 142, pp. 71-128.
Hadjichristidis, N. see *Xu, Z.*: Vol. 120, pp. 1-50.
Hadjichristidis, N. see *Pitsikalis, M.*: Vol. 135, pp. 1-138.
Hahn, O. see *Baschnagel, J.*: Vol. 152, pp. 41-156.
Hall, H. K. see *Penelle, J.*: Vol. 102, pp. 73-104.
Höcker, H. see *Klee, D.*: Vol. 149, pp. 1-57.
Hammouda, B.: SANS from Homogeneous Polymer Mixtures: A Unified Overview. Vol. 106, pp. 87-134.
Han, M.J. and *Chang, J.Y.*: Polynucleotide Analogues. Vol. 153, pp. 1-36.
Harada, A.: Design and Construction of Supramolecular Architectures Consisting of Cyclodextrins and Polymers. Vol. 133, pp. 141-192.
Haralson, M. A. see *Prokop, A.*: Vol. 136, pp. 1-52.

Hassan, C.M. and Peppas, N.A.: Structure and Applications of Poly(vinyl alcohol) Hydrogels Produced by Conventional Crosslinking or by Freezing/Thawing Methods. Vol. 153, pp. 37-65.
Hawker, C. J. Dentritic and Hyperbranched Macromolecules – Precisely Controlled Macromolecular Architectures. Vol. 147, pp. 113-160.
Hawker, C. J. see Hedrick, J. L.: Vol. 141, pp. 1-44.
Hedrick, J. L., Carter, K. R., Labadie, J. W., Miller, R. D., Volksen, W., Hawker, C. J., Yoon, D. Y., Russell, T. P., McGrath, J. E., Briber, R. M.: Nanoporous Polyimides. Vol. 141, pp. 1-44.
Hedrick, J. L., Labadie, J. W., Volksen, W. and *Hilborn, J. G.*: Nanoscopically Engineered Polyimides. Vol. 147, pp. 61-112.
Hedrick, J. L. see Hergenrother, P. M.: Vol. 117, pp. 67-110.
Hedrick, J. L. see Kiefer, J.: Vol. 147, pp. 161-247.
Hedrick, J.L. see McGrath, J. E.: Vol. 140, pp. 61-106.
Heller, J.: Poly (Ortho Esters). Vol. 107, pp. 41-92.
Hemielec, A. A. see Hunkeler, D.: Vol. 112, pp. 115-134.
Hergenrother, P. M., Connell, J. W., Labadie, J. W. and *Hedrick, J. L.*: Poly(arylene ether)s Containing Heterocyclic Units. Vol. 117, pp. 67-110.
Hernández-Barajas, J. see Wandrey, C.: Vol. 145, pp. 123-182.
Hervet, H. see Léger, L.: Vol. 138, pp. 185-226.
Hilborn, J. G. see Hedrick, J. L.: Vol. 147, pp. 61-112.
Hilborn, J. G. see Kiefer, J.: Vol. 147, pp. 161-247.
Hiramatsu, N. see Matsushige, M.: Vol. 125, pp. 147-186.
Hirasa, O. see Suzuki, M.: Vol. 110, pp. 241-262.
Hirotsu, S.: Coexistence of Phases and the Nature of First-Order Transition in Poly-N-isopropylacrylamide Gels. Vol. 110, pp. 1-26.
Hamley, I. W.: Crystallization in Block Copolymers. Vol. 148, pp. 113-138.
Hornsby, P.: Rheology, Compoundind and Processing of Filled Thermoplastics. Vol. 139, pp. 155-216.
Hui, C.-Y. see Creton, C.: Vol. 156, pp. 53-135
Hult, A., Johansson, M., Malmström, E.: Hyperbranched Polymers. Vol. 143, pp. 1-34.
Hunkeler, D., Candau, F., Pichot, C., Hemielec, A. E., Xie, T. Y., Barton, J., Vaskova, V., Guillot, J., Dimonie, M. V., Reichert, K. H.: Heterophase Polymerization: A Physical and Kinetic Comparision and Categorization. Vol. 112, pp. 115-134.
Hunkeler, D. see Prokop, A.: Vol. 136, pp. 1-52; 53-74.
Hunkeler, D see Wandrey, C.: Vol. 145, pp. 123-182.

Iatrou, H. see Hadjichristidis, N.: Vol. 142, pp. 71-128.
Ichikawa, T. see Yoshida, H.: Vol. 105, pp. 3-36.
Ihara, E. see Yasuda, H.: Vol. 133, pp. 53-102.
Ikada, Y. see Uyama,Y.: Vol. 137, pp. 1-40.
Ilavsky, M.: Effect on Phase Transition on Swelling and Mechanical Behavior of Synthetic Hydrogels. Vol. 109, pp. 173-206.
Imai, Y.: Rapid Synthesis of Polyimides from Nylon-Salt Monomers. Vol. 140, pp. 1-23.
Inomata, H. see Saito, S.: Vol. 106, pp. 207-232.
Inoue, S. see Sugimoto, H.: Vol. 146, pp. 39-120.
Irie, M.: Stimuli-Responsive Poly(N-isopropylacrylamide), Photo- and Chemical-Induced Phase Transitions. Vol. 110, pp. 49-66.
Ise, N. see Matsuoka, H.: Vol. 114, pp. 187-232.
*Ito, K., Kawaguchi, S,:*Poly(macronomers), Homo- and Copolymerization. Vol. 142, pp. 129-178.
Ivanov, A. E. see Zubov, V. P.: Vol. 104, pp. 135-176.

Jacob, S. and Kennedy, J.: Synthesis, Characterization and Properties of OCTA-ARM Polyisobutylene-Based Star Polymers. Vol. 146, pp. 1-38.
Jaffe, M., Chen, P., Choe, E.-W., Chung, T.-S. and *Makhija, S.*: High Performance Polymer Blends. Vol. 117, pp. 297-328.
Jancar, J.: Structure-Property Relationships in Thermoplastic Matrices. Vol. 139, pp. 1-66.

Jerôme, R.: see Mecerreyes, D.: Vol. 147, pp. 1-60.
Jiang, M., Li, M., Xiang, M. and *Zhou, H.*: Interpolymer Complexation and Miscibility and Enhancement by Hydrogen Bonding. Vol. 146, pp. 121-194.
Jo, W. H. and *Yang, J. S.*: Molecular Simulation Approaches for Multiphase Polymer Systems. Vol. 156, pp. 1-52.
Johansson, M. see Hult, A.: Vol. 143, pp. 1-34.
Joos-Müller, B. see Funke, W.: Vol. 136, pp. 137-232.
Jou, D., Casas-Vazquez, J. and *Criado-Sancho, M.*: Thermodynamics of Polymer Solutions under Flow: Phase Separation and Polymer Degradation. Vol. 120, pp. 207-266.

Kaetsu, I.: Radiation Synthesis of Polymeric Materials for Biomedical and Biochemical Applications. Vol. 105, pp. 81-98.
Kaji, K. see Kanaya, T.: Vol. 154, pp. 87-141.
Kakimoto, M. see Gaw, K. O.: Vol. 140, pp. 107-136.
Kaminski, W. and *Arndt, M.*: Metallocenes for Polymer Catalysis. Vol. 127, pp. 143-187.
Kammer, H. W., Kressler, H. and *Kummerloewe, C.*: Phase Behavior of Polymer Blends - Effects of Thermodynamics and Rheology. Vol. 106, pp. 31-86.
Kanaya, T. and *Kaji, K.*: Dynamcis in the Glassy State and Near the Glass Transition of Amorphous Polymers as Studied by Neutron Scattering. Vol. 154, pp. 87-141.
Kandyrin, L. B. and *Kuleznev, V. N.*: The Dependence of Viscosity on the Composition of Concentrated Dispersions and the Free Volume Concept of Disperse Systems. Vol. 103, pp. 103-148.
Kaneko, M. see Ramaraj, R.: Vol. 123, pp. 215-242.
Kang, E. T., Neoh, K. G. and *Tan, K. L.*: X-Ray Photoelectron Spectroscopic Studies of Electroactive Polymers. Vol. 106, pp. 135-190.
Kato, K. see Uyama, Y.: Vol. 137, pp. 1-40.
Kawaguchi, S. see Ito, K.: Vol. 142, p 129-178.
Kazanskii, K. S. and *Dubrovskii, S. A.*: Chemistry and Physics of „Agricultural" Hydrogels. Vol. 104, pp. 97-134.
Kennedy, J. P. see Jacob, S.: Vol. 146, pp. 1-38.
Kennedy, J. P. see Majoros, I.: Vol. 112, pp. 1-113.
Khokhlov, A., Starodybtzev, S. and *Vasilevskaya, V.*: Conformational Transitions of Polymer Gels: Theory and Experiment. Vol. 109, pp. 121-172.
Kiefer, J., Hedrick J. L. and *Hiborn, J. G.*: Macroporous Thermosets by Chemically Induced Phase Separation. Vol. 147, pp. 161-247.
Kilian, H. G. and *Pieper, T.*: Packing of Chain Segments. A Method for Describing X-Ray Patterns of Crystalline, Liquid Crystalline and Non-Crystalline Polymers. Vol. 108, pp. 49-90.
Kim, J. see Quirk, R.P.: Vol. 153, pp. 67-162.
Kishore, K. and *Ganesh, K.*: Polymers Containing Disulfide, Tetrasulfide, Diselenide and Ditelluride Linkages in the Main Chain. Vol. 121, pp. 81-122.
Kitamaru, R.: Phase Structure of Polyethylene and Other Crystalline Polymers by Solid-State ^{13}C/MNR. Vol. 137, pp 41-102.
Klee, D. and *Höcker, H.*: Polymers for Biomedical Applications: Improvement of the Interface Compatibility. Vol. 149, pp. 1-57.
Klier, J. see Scranton, A. B.: Vol. 122, pp. 1-54.
Kobayashi, S., Shoda, S. and *Uyama, H.*: Enzymatic Polymerization and Oligomerization. Vol. 121, pp. 1-30.
Köhler, W. and *Schäfer, R.*: Polymer Analysis by Thermal-Diffusion Forced Rayleigh Scattering. Vol. 151, pp. 1-59.
Koenig, J. L. see Andreis, M.: Vol. 124, pp. 191-238.
Koike, T.: Viscoelastic Behavior of Epoxy Resins Before Crosslinking. Vol. 148, pp. 139-188.
Kokufuta, E.: Novel Applications for Stimulus-Sensitive Polymer Gels in the Preparation of Functional Immobilized Biocatalysts. Vol. 110, pp. 157-178.
Konno, M. see Saito, S.: Vol. 109, pp. 207-232.

Kopecek, J. see Putnam, D.: Vol. 122, pp. 55-124.
Koßmehl, G. see Schopf, G.: Vol. 129, pp. 1-145.
Kramer, E. J. see Creton, C.: Vol. 156, pp. 53-135.
Kremer, K. see Baschnagel, J.: Vol. 152, pp. 41-156.
Kressler, J. see Kammer, H. W.: Vol. 106, pp. 31-86.
Kricheldorf, H. R.: Liquid-Cristalline Polyimides. Vol. 141, pp. 83-188.
Krishnamoorti, R. see Giannelis, E.P.: Vol. 138, pp. 107-148.
Kirchhoff, R. A. and *Bruza, K. J.*: Polymers from Benzocyclobutenes. Vol. 117, pp. 1-66.
Kuchanov, S. I.: Modern Aspects of Quantitative Theory of Free-Radical Copolymerization. Vol. 103, pp. 1-102.
Kuchanov, S. I.: Principles of Quantitative Description of Chemical Structure of Synthetic Polymers. Vol. 152, p. 157-202.
Kudaibergennow, S.E.: Recent Advances in Studying of Synthetic Polyampholytes in Solutions. Vol. 144, pp. 115-198.
Kuleznev, V. N. see Kandyrin, L. B.: Vol. 103, pp. 103-148.
Kulichkhin, S. G. see Malkin, A. Y.: Vol. 101, pp. 217-258.
Kulicke, W.-M. see Grigorescu, G.: Vol. 152, p. 1-40.
Kummerloewe, C. see Kammer, H. W.: Vol. 106, pp. 31-86.
Kuznetsova, N. P. see Samsonov, G. V.: Vol. 104, pp. 1-50. Labadie, J. W. see Hergenrother, P. M.: Vol. 117, pp. 67-110.

Labadie, J. W. see Hedrick, J. L.: Vol. 141, pp. 1-44.
Labadie, J. W. see Hedrick, J. L.: Vol. 147, pp. 61-112.
Lamparski, H. G. see O´Brien, D. F.: Vol. 126, pp. 53-84.
Laschewsky, A.: Molecular Concepts, Self-Organisation and Properties of Polysoaps. Vol. 124, pp. 1-86.
Laso, M. see Leontidis, E.: Vol. 116, pp. 283-318.
Lazár, M. and *RychlΩ, R.*: Oxidation of Hydrocarbon Polymers. Vol. 102, pp. 189-222.
Lechowicz, J. see Galina, H.: Vol. 137, pp. 135-172.
Léger, L., Raphaël, E., Hervet, H.: Surface-Anchored Polymer Chains: Their Role in Adhesion and Friction. Vol. 138, pp. 185-226.
Lenz, R. W.: Biodegradable Polymers. Vol. 107, pp. 1-40.
Leontidis, E., de Pablo, J. J., Laso, M. and *Suter, U. W.*: A Critical Evaluation of Novel Algorithms for the Off-Lattice Monte Carlo Simulation of Condensed Polymer Phases. Vol. 116, pp. 283-318.
Lee, B. see Quirk, R.P: Vol. 153, pp. 67-162.
Lee, Y. see Quirk, R.P: Vol. 153, pp. 67-162.
Lesec, J. see Viovy, J.-L.: Vol. 114, pp. 1-42.
Li, M. see Jiang, M.: Vol. 146, pp. 121-194.
Liang, G. L. see Sumpter, B. G.: Vol. 116, pp. 27-72.
Lienert, K.-W.: Poly(ester-imide)s for Industrial Use. Vol. 141, pp. 45-82.
Lin, J. and *Sherrington, D. C.*: Recent Developments in the Synthesis, Thermostability and Liquid Crystal Properties of Aromatic Polyamides. Vol. 111, pp. 177-220.
López Cabarcos, E. see Baltá-Calleja, F. J.: Vol. 108, pp. 1-48.

Majoros, I., Nagy, A. and *Kennedy, J. P.*: Conventional and Living Carbocationic Polymerizations United. I. A Comprehensive Model and New Diagnostic Method to Probe the Mechanism of Homopolymerizations. Vol. 112, pp. 1-113.
Makhija, S. see Jaffe, M.: Vol. 117, pp. 297-328.
Malmström, E. see Hult, A.: Vol. 143, pp. 1-34.
Malkin, A. Y. and *Kulichkhin, S. G.*: Rheokinetics of Curing. Vol. 101, pp. 217-258.
Maniar, M. see Domb, A. J.: Vol. 107, pp. 93-142.
Manias, E., see Giannelis, E.P.: Vol. 138, pp. 107-148.
Mashima, K., Nakayama, Y. and *Nakamura, A.*: Recent Trends in Polymerization of a-Olefins Catalyzed by Organometallic Complexes of Early Transition Metals. Vol. 133, pp. 1-52.

Mathew, D. see Reghunadhan Nair, C.P.: Vol. 155, pp. 1-99.
Matsumoto, A.: Free-Radical Crosslinking Polymerization and Copolymerization of Multivinyl Compounds. Vol. 123, pp. 41-80.
Matsumoto, A. see Otsu, T.: Vol. 136, pp. 75-138.
Matsuoka, H. and *Ise, N.*: Small-Angle and Ultra-Small Angle Scattering Study of the Ordered Structure in Polyelectrolyte Solutions and Colloidal Dispersions. Vol. 114, pp. 187-232.
Matsushige, K., Hiramatsu, N. and *Okabe, H.*: Ultrasonic Spectroscopy for Polymeric Materials. Vol. 125, pp. 147-186.
Mattice, W. L. see Rehahn, M.: Vol. 131/132, pp. 1-475.
Mattice, W. L. see Baschnagel, J.: Vol. 152, p. 41-156.
Mays, W. see Xu, Z.: Vol. 120, pp. 1-50.
Mays, J.W. see Pitsikalis, M.: Vol.135, pp. 1-138.
McGrath, J. E. see Hedrick, J. L.: Vol. 141, pp. 1-44.
McGrath, J. E., Dunson, D. L., Hedrick, J. L.: Synthesis and Characterization of Segmented Polyimide-Polyorganosiloxane Copolymers. Vol. 140, pp. 61-106.
McLeish, T.C.B., Milner, S. T.: Entangled Dynamics and Melt Flow of Branched Polymers. Vol. 143, pp. 195-256.
Mecerreyes, D., Dubois, P. and *Jerôme, R.*: Novel Macromolecular Architectures Based on Aliphatic Polyesters: Relevance of the „Coordination-Insertion" Ring-Opening Polymerization. Vol. 147, pp. 1 -60.
Mecham, S. J. see McGrath, J. E.: Vol. 140, pp. 61-106.
Mikos, A. G. see Thomson, R. C.: Vol. 122, pp. 245-274.
Milner, S. T. see McLeish, T. C. B.: Vol. 143, pp. 195-256.
Mison, P. and *Sillion, B.*: Thermosetting Oligomers Containing Maleimides and Nadiimides End-Groups. Vol. 140, pp. 137-180.
Miyasaka, K.: PVA-Iodine Complexes: Formation, Structure and Properties. Vol. 108. pp. 91-130.
Miller, R. D. see Hedrick, J. L.: Vol. 141, pp. 1-44.
Monnerie, L. see Bahar, I.: Vol. 116, pp. 145-206.
Morishima, Y.: Photoinduced Electron Transfer in Amphiphilic Polyelectrolyte Systems. Vol. 104, pp. 51-96.
Morton M. see Quirk, R.P: Vol. 153, pp. 67-162
Mours, M. see Winter, H. H.: Vol. 134, pp. 165-234.
Müllen, K. see Scherf, U.: Vol. 123, pp. 1-40.
Müller-Plathe, F. see Gusev, A. A.: Vol. 116, pp. 207-248.
Müller-Plathe, F. see Baschnagel, J.: Vol. 152, p. 41-156.
Mukerherjee, A. see Biswas, M.: Vol. 115, pp. 89-124.
Murat, M. see Baschnagel, J.: Vol. 152, p. 41-156.
Mylnikov, V.: Photoconducting Polymers. Vol. 115, pp. 1-88.

Nagy, A. see Majoros, I.: Vol. 112, pp. 1-11.
Nakamura, A. see Mashima, K.: Vol. 133, pp. 1-52.
Nakayama, Y. see Mashima, K.: Vol. 133, pp. 1-52.
Narasinham, B., Peppas, N. A.: The Physics of Polymer Dissolution: Modeling Approaches and Experimental Behavior. Vol. 128, pp. 157-208.
Nechaev, S. see Grosberg, A.: Vol. 106, pp. 1-30.
Neoh, K. G. see Kang, E. T.: Vol. 106, pp. 135-190.
Newman, S. M. see Anseth, K. S.: Vol. 122, pp. 177-218.
Nijenhuis, K. te: Thermoreversible Networks. Vol. 130, pp. 1-252.
Ninan, K.N. see Reghunadhan Nair, C.P.: Vol. 155, pp. 1-99.
Noid, D. W. see Otaigbe, J.U.: Vol. 154, pp. 1-86.
Noid, D. W. see Sumpter, B. G.: Vol. 116, pp. 27-72.
Novac, B. see Grubbs, R.: Vol. 102, pp. 47-72.
Novikov, V. V. see Privalko, V. P.: Vol. 119, pp. 31-78.

O'Brien, D. F., Armitage, B. A., Bennett, D. E. and *Lamparski, H. G.*: Polymerization and Domain Formation in Lipid Assemblies. Vol. 126, pp. 53-84.

Ogasawara, M.: Application of Pulse Radiolysis to the Study of Polymers and Polymerizations. Vol.105, pp. 37-80.
Okabe, H. see Matsushige, K.: Vol. 125, pp. 147-186.
Okada, M.: Ring-Opening Polymerization of Bicyclic and Spiro Compounds. Reactivities and Polymerization Mechanisms. Vol. 102, pp. 1-46.
Okano, T.: Molecular Design of Temperature-Responsive Polymers as Intelligent Materials. Vol. 110, pp. 179-198.
Okay, O. see Funke, W.: Vol. 136, pp. 137-232.
Onuki, A.: Theory of Phase Transition in Polymer Gels. Vol. 109, pp. 63-120.
Osad'ko, I.S.: Selective Spectroscopy of Chromophore Doped Polymers and Glasses. Vol. 114, pp. 123-186.
Otaigbe, J. U., Barnes, M. D., Fukui, K., Sumpter, B. G., Noid, D. W.: Generation, Characterization, and Modeling of Polymer Micro- and Nano-Particles. Vol. 154, pp. 1-86.
Otsu, T., Matsumoto, A.: Controlled Synthesis of Polymers Using the Iniferter Technique: Developments in Living Radical Polymerization. Vol. 136, pp. 75-138.

de Pablo, J. J. see Leontidis, E.: Vol. 116, pp. 283-318.
Padias, A. B. see Penelle, J.: Vol. 102, pp. 73-104.
Pascault, J.-P. see Williams, R. J. J.: Vol. 128, pp. 95-156.
Pasch, H.: Analysis of Complex Polymers by Interaction Chromatography. Vol. 128, pp. 1-46.
Pasch, H.: Hyphenated Techniques in Liquid Chromatography of Polymers. Vol. 150, pp. 1-66.
Paul, W. see Baschnagel, J.: Vol. 152, p. 41-156.
Penczek, P. see Batog, A. E.: Vol. 144, pp. 49-114.
Penelle, J., Hall, H. K., Padias, A. B. and *Tanaka, H.:* Captodative Olefins in Polymer Chemistry. Vol. 102, pp. 73-104.
Peppas, N. A. see Bell, C. L.: Vol. 122, pp. 125-176.
Peppas, N.A. see Hassan, C.M.: Vol. 153, pp. 37-65
Peppas, N. A. see Narasimhan, B.: Vol. 128, pp. 157-208.
Pet'ko, I. P. see Batog, A. E.: Vol. 144, pp. 49-114.
Pichot, C. see Hunkeler, D.: Vol. 112, pp. 115-134.
Pieper, T. see Kilian, H. G.: Vol. 108, pp. 49-90.
Pispas, S. see Pitsikalis, M.: Vol. 135, pp. 1-138.
Pispas, S. see Hadjichristidis: Vol. 142, pp. 71-128.
Pitsikalis, M., Pispas, S., Mays, J. W., Hadjichristidis, N.: Nonlinear Block Copolymer Architectures. Vol. 135, pp. 1-138.
Pitsikalis, M. see Hadjichristidis: Vol. 142, pp. 71-128.
Pötschke, D. see Dingenouts, N.: Vol 144, pp. 1-48.
Pokrovskii, V. N.: The Mesoscopic Theory of the Slow Relaxation of Linear Macromolecules. Vol. 154, pp. 143-219.
Pospíšil, J.: Functionalized Oligomers and Polymers as Stabilizers for Conventional Polymers. Vol. 101, pp. 65-168.
Pospíšil, J.: Aromatic and Heterocyclic Amines in Polymer Stabilization. Vol. 124, pp. 87-190.
Powers, A. C. see Prokop, A.: Vol. 136, pp. 53-74.
Priddy, D. B.: Recent Advances in Styrene Polymerization. Vol. 111, pp. 67-114.
Priddy, D. B.: Thermal Discoloration Chemistry of Styrene-co-Acrylonitrile. Vol. 121, pp. 123-154.
Privalko, V. P. and *Novikov, V. V.:* Model Treatments of the Heat Conductivity of Heterogeneous Polymers. Vol. 119, pp 31-78.
Prokop, A., Hunkeler, D., Powers, A. C., Whitesell, R. R., Wang, T. G.: Water Soluble Polymers for Immunoisolation II: Evaluation of Multicomponent Microencapsulation Systems. Vol. 136, pp. 53-74.
Prokop, A., Hunkeler, D., DiMari, S., Haralson, M. A., Wang, T. G.: Water Soluble Polymers for Immunoisolation I: Complex Coacervation and Cytotoxicity. Vol. 136, pp. 1-52.
Pukánszky, B. and *Fekete, E.:* Adhesion and Surface Modification. Vol. 139, pp. 109-154.
Putnam, D. and *Kopecek, J.:* Polymer Conjugates with Anticancer Acitivity. Vol. 122, pp. 55-124.

Quirk, R.P. and Yoo, T., Lee, Y., M., Kim, J. and Lee, B.: Applications of 1,1-Diphenylethylene Chemistry in Anionic Synthesis of Polymers with Controlled Structures. Vol. 153, pp. 67-162.

Ramaraj, R. and Kaneko, M.: Metal Complex in Polymer Membrane as a Model for Photosynthetic Oxygen Evolving Center. Vol. 123, pp. 215-242.
Rangarajan, B. see Scranton, A. B.: Vol. 122, pp. 1-54.
Raphaël, E. see Léger, L.: Vol. 138, pp. 185-226.
Reddinger, J. L. and Reynolds, J. R.: Molecular Engineering of π-Conjugated Polymers. Vol. 145, pp. 57-122.
Reghunadhan Nair, C.P., Mathew, D. and Ninan, K.N., : Cyanate Ester Resins, Recent Developments. Vol. 155, pp. 1-99.
Reichert, K. H. see Hunkeler, D.: Vol. 112, pp. 115-134.
Rehahn, M., Mattice, W. L., Suter, U. W.: Rotational Isomeric State Models in Macromolecular Systems. Vol. 131/132, pp. 1-475.
Reynolds, J.R. see Reddinger, J. L.: Vol. 145, pp. 57-122.
Richter, D. see Ewen, B.: Vol. 134, pp.1-130.
Risse, W. see Grubbs, R.: Vol. 102, pp. 47-72.
Rivas, B. L. and Geckeler, K. E.: Synthesis and Metal Complexation of Poly(ethyleneimine) and Derivatives. Vol. 102, pp. 171-188.
Robin, J. J. see Boutevin, B.: Vol. 102, pp. 105-132.
Roe, R.-J.: MD Simulation Study of Glass Transition and Short Time Dynamics in Polymer Liquids. Vol. 116, pp. 111-114.
Roovers, J., Comanita, B.: Dendrimers and Dendrimer-Polymer Hybrids. Vol. 142, pp 179-228.
Rothon, R. N.: Mineral Fillers in Thermoplastics: Filler Manufacture and Characterisation. Vol. 139, pp. 67-108.
Rozenberg, B. A. see Williams, R. J. J.: Vol. 128, pp. 95-156.
Ruckenstein, E.: Concentrated Emulsion Polymerization. Vol. 127, pp. 1-58.
Rusanov, A. L.: Novel Bis (Naphtalic Anhydrides) and Their Polyheteroarylenes with Improved Processability. Vol. 111, pp. 115-176.
Russel, T. P. see Hedrick, J. L.: Vol. 141, pp. 1-44.
Rychlý, J. see Lazár, M.: Vol. 102, pp. 189-222.
Ryzhov, V. A. see Bershtein, V. A.: Vol. 114, pp. 43-122.

Sabsai, O. Y. see Barshtein, G. R.: Vol. 101, pp. 1-28.
Saburov, V. V. see Zubov, V. P.: Vol. 104, pp. 135-176.
Saito, S., Konno, M. and Inomata, H.: Volume Phase Transition of N-Alkylacrylamide Gels. Vol. 109, pp. 207-232.
Samsonov, G. V. and Kuznetsova, N. P.: Crosslinked Polyelectrolytes in Biology. Vol. 104, pp. 1-50.
Santa Cruz, C. see Baltá-Calleja, F. J.: Vol. 108, pp. 1-48.
Santos, S. see Baschnagel, J.: Vol. 152, p. 41-156.
Sato, T. and Teramoto, A.: Concentrated Solutions of Liquid-Christalline Polymers. Vol. 126, pp. 85-162.
Schäfer R. see Köhler, W.: Vol. 151, pp. 1-59.
Scherf, U. and Müllen, K.: The Synthesis of Ladder Polymers. Vol. 123, pp. 1-40.
Schmidt, M. see Förster, S.: Vol. 120, pp. 51-134.
Schopf, G. and Koßmehl, G.: Polythiophenes - Electrically Conductive Polymers. Vol. 129, pp. 1-145.
Schweizer, K. S.: Prism Theory of the Structure, Thermodynamics, and Phase Transitions of Polymer Liquids and Alloys. Vol. 116, pp. 319-378.
Scranton, A. B., Rangarajan, B. and Klier, J.: Biomedical Applications of Polyelectrolytes. Vol. 122, pp. 1-54.
Sefton, M. V. and Stevenson, W. T. K.: Microencapsulation of Live Animal Cells Using Polycrylates. Vol.107, pp. 143-198.

Shamanin, V. V.: Bases of the Axiomatic Theory of Addition Polymerization. Vol. 112, pp. 135-180.
Sheiko, S. S.: Imaging of Polymers Using Scanning Force Microscopy: From Superstructures to Individual Molecules. Vol. 151, pp. 61-174.
Sherrington, D. C. see Cameron, N. R., Vol. 126, pp. 163-214.
Sherrington, D. C. see Lin, J.: Vol. 111, pp. 177-220.
Sherrington, D. C. see Steinke, J.: Vol. 123, pp. 81-126.
Shibayama, M. see Tanaka, T.: Vol. 109, pp. 1-62.
Shiga, T.: Deformation and Viscoelastic Behavior of Polymer Gels in Electric Fields. Vol. 134, pp. 131-164.
Shoda, S. see Kobayashi, S.: Vol. 121, pp. 1-30.
Siegel, R. A.: Hydrophobic Weak Polyelectrolyte Gels: Studies of Swelling Equilibria and Kinetics. Vol. 109, pp. 233-268.
Silvestre, F. see Calmon-Decriaud, A.: Vol. 207, pp. 207-226.
Sillion, B. see Mison, P.: Vol. 140, pp. 137-180.
Singh, R. P. see Sivaram, S.: Vol. 101, pp. 169-216.
Sinha Ray, S. see Biswas, M: Vol. 155, pp. 167-221.
Sivaram, S. and *Singh, R. P.*: Degradation and Stabilization of Ethylene-Propylene Copolymers and Their Blends: A Critical Review. Vol. 101, pp. 169-216.
Starodybtzev, S. see Khokhlov, A.: Vol. 109, pp. 121-172.
Steinke, J., Sherrington, D. C. and *Dunkin, I. R.*: Imprinting of Synthetic Polymers Using Molecular Templates. Vol. 123, pp. 81-126.
Stenzenberger, H. D.: Addition Polyimides. Vol. 117, pp. 165-220.
Stevenson, W. T. K. see Sefton, M. V.: Vol. 107, pp. 143-198.
Suematsu, K.: Recent Progress of Gel Theory: Ring, Excluded Volume, and Dimension. Vol. 156, pp. 136-214.
Sumpter, B. G., Noid, D. W., Liang, G. L. and *Wunderlich, B.*: Atomistic Dynamics of Macromolecular Crystals. Vol. 116, pp. 27-72.
Sumpter, B. G. see Otaigbe, J.U.: Vol. 154, pp. 1-86.
Sugimoto, H. and *Inoue, S.*: Polymerization by Metalloporphyrin and Related Complexes. Vol. 146, pp. 39-120.
Suter, U. W. see Gusev, A. A.: Vol. 116, pp. 207-248.
Suter, U. W. see Leontidis, E.: Vol. 116, pp. 283-318.
Suter, U. W. see Rehahn, M.: Vol. 131/132, pp. 1-475.
Suter, U. W. see Baschnagel, J.: Vol. 152, p. 41-156.
Suzuki, A.: Phase Transition in Gels of Sub-Millimeter Size Induced by Interaction with Stimuli. Vol. 110, pp. 199-240.
Suzuki, A. and *Hirasa, O.*: An Approach to Artifical Muscle by Polymer Gels due to Micro-Phase Separation. Vol. 110, pp. 241-262.

Tagawa, S.: Radiation Effects on Ion Beams on Polymers. Vol. 105, pp. 99-116.
Tan, K. L. see Kang, E. T.: Vol. 106, pp. 135-190.
Tanaka, T. see Penelle, J.: Vol. 102, pp. 73-104.
Tanaka, H. and *Shibayama, M.*: Phase Transition and Related Phenomena of Polymer Gels. Vol. 109, pp. 1-62.
Tauer, K. see Guyot, A.: Vol. 111, pp. 43-66.
Teramoto, A. see Sato, T.: Vol. 126, pp. 85-162.
Terent'eva, J. P. and *Fridman, M. L.*: Compositions Based on Aminoresins. Vol. 101, pp. 29-64.
Theodorou, D. N. see Dodd, L. R.: Vol. 116, pp. 249-282.
Thomson, R. C., Wake, M. C., Yaszemski, M. J. and *Mikos, A. G.*: Biodegradable Polymer Scaffolds to Regenerate Organs. Vol. 122, pp. 245-274.
Tokita, M.: Friction Between Polymer Networks of Gels and Solvent. Vol. 110, pp. 27-48.
Tries, V. see Baschnagel, J:. Vol. 152, p. 41-156.
Tsuruta, T.: Contemporary Topics in Polymeric Materials for Biomedical Applications. Vol. 126, pp. 1-52.

Uyama, H. see Kobayashi, S.: Vol. 121, pp. 1-30.
Uyama, Y: Surface Modification of Polymers by Grafting. Vol. 137, pp. 1-40.

Vasilevskaya, V. see Khokhlov, A.: Vol. 109, pp. 121-172.
Vaskova, V. see Hunkeler, D.: Vol.:112, pp. 115-134.
Verdugo, P.: Polymer Gel Phase Transition in Condensation-Decondensation of Secretory Products. Vol. 110, pp. 145-156.
Vettegren, V. I.: see Bronnikov, S. V.: Vol. 125, pp. 103-146.
Viovy, J.-L. and *Lesec, J.*: Separation of Macromolecules in Gels: Permeation Chromatography and Electrophoresis. Vol. 114, pp. 1-42.
Vlahos, C. see Hadjichristidis, N.: Vol. 142, pp. 71-128.
Volksen, W.: Condensation Polyimides: Synthesis, Solution Behavior, and Imidization Characteristics. Vol. 117, pp. 111-164.
Volksen, W. see Hedrick, J. L.: Vol. 141, pp. 1-44.
Volksen, W. see Hedrick, J. L.: Vol. 147, pp. 61-112.

Wake, M. C. see Thomson, R. C.: Vol. 122, pp. 245-274.
Wandrey C., Hernández-Barajas, J. and *Hunkeler, D.*: Diallyldimethylammonium Chloride and its Polymers. Vol. 145, pp. 123-182.
Wang, K. L. see Cussler, E. L.: Vol. 110, pp. 67-80.
Wang, S.-Q.: Molecular Transitions and Dynamics at Polymer/Wall Interfaces: Origins of Flow Instabilities and Wall Slip. Vol. 138, pp. 227-276.
Wang, T. G. see Prokop, A.: Vol. 136, pp.1-52; 53-74.
Whitesell, R. R. see Prokop, A.: Vol. 136, pp. 53-74.
Williams, R. J. J., Rozenberg, B. A., Pascault, J.-P.: Reaction Induced Phase Separation in Modified Thermosetting Polymers. Vol. 128, pp. 95-156.
Winter, H. H., Mours, M.: Rheology of Polymers Near Liquid-Solid Transitions. Vol. 134, pp. 165-234.
Wu, C.: Laser Light Scattering Characterization of Special Intractable Macromolecules in Solution. Vol 137, pp. 103-134.
Wunderlich, B. see Sumpter, B. G.: Vol. 116, pp. 27-72.

Xiang, M. see Jiang, M.: Vol. 146, pp. 121-194.
Xie, T. Y. see Hunkeler, D.: Vol. 112, pp. 115-134.
Xu, Z., Hadjichristidis, N., Fetters, L. J. and *Mays, J. W.*: Structure/Chain-Flexibility Relationships of Polymers. Vol. 120, pp. 1-50.

Yagci, Y. and *Endo, T.*: N-Benzyl and N-Alkoxy Pyridium Salts as Thermal and Photochemical Initiators for Cationic Polymerization. Vol. 127, pp. 59-86.
Yannas, I. V.: Tissue Regeneration Templates Based on Collagen-Glycosaminoglycan Copolymers. Vol. 122, pp. 219-244.
Yang, J. S. see Jo, W. H.: Vol. 156, pp. 1-52.
Yamaoka, H.: Polymer Materials for Fusion Reactors. Vol. 105, pp. 117-144.
Yasuda, H. and *Ihara, E.*: Rare Earth Metal-Initiated Living Polymerizations of Polar and Nonpolar Monomers. Vol. 133, pp. 53-102.
Yaszemski, M. J. see Thomson, R. C.: Vol. 122, pp. 245-274.
Yoo, T. see Quirk, R.P.: Vol. 153, pp. 67-162.
Yoon, D. Y. see Hedrick, J. L.: Vol. 141, pp. 1-44.
Yoshida, H. and *Ichikawa, T.*: Electron Spin Studies of Free Radicals in Irradiated Polymers. Vol. 105, pp. 3-36.

Zhou, H. see Jiang, M.: Vol. 146, pp. 121-194.
Zubov, V. P., Ivanov, A. E. and *Saburov, V. V.*: Polymer-Coated Adsorbents for the Separation of Biopolymers and Particles. Vol. 104, pp. 135-176.

Subject Index

abnormality in chemical machinery (177–178), 193
abnormality in critical behavior 193
analytical expression of gel point 188
approximate solution of ring concentration 188
asymptotic equation 163–165, 169
average profile 145

basic equality 186, 193, 210
Bethe lattice 173, 178
bond animal 152, 181, 207
bond percolation threshold 205-208
bond transition 154
boundary concentration, C^* 164-165
boundary dimensionality, d^* 173
branched molecule 150–152, 154, 168
branching process 143–146, 154, 177
bifunctional system 163–164

cascade formalism 143
Cauchy theorem 182
classical theory of gelation 143
concentration invariance 164, 165, 167
critical dilution, γ_c 189, 194, 199, 201–203
critical dimension, d_c 149, 154, 210
crossover point 178–179
cyclic bond 184
cyclization on lattice 168
cyclization probability, $\mathcal{P}$ 156
cyclization rate 155, 157, 170

d'Alembert theorem 181
des Cloizeaux type equation 152
Desmodur RF 201-202
d-dimensional sphere 148, 156, 172, 175
dialdehyde as cross-linker 142
dilution limit 151–152, 206, 213
dimension 149, 154, 158

elastic solid 178
end-to-end distance 147
entropy term 149, 172, 177

excess bonds 184
excluded volume 146, 171, 196 (footnote), 210, 213
excluded volume parameter, v 148–149
expansion factor, α 151
extent of reaction of intermolecular reaction, D(inter) 184–185, 209
extent of reaction of cyclization, D(ring) 184–185
extra factor, ω_j 158, 161

Flory excluded volume theory 151, 169
Flory-Fisher theory 149
fundamental equality 155

Gaussian chain 153
Gaussian statistics 147, 149
gel time 179
gem-substituent effect 197–198
generation 144–145, 157, 159, 170

high concentration expansion 187
high dimension expansion 191, 192, 203
hyperbolic relation 191
hypercubic lattice 172, 204, 209

incomplete Gamma function 156, 205
infinite concentration 146, 155, 195 (footnote)
intermolecular reaction 157, 171, 177
Isaacson-Lubensky formula 152

Le Chatelier's law 189
limiting case 173
limiting solution 163–165, 169, 174
linear relationship 191
local potential energy 148
loss tangent 178–179

marginal dimensionality 152, 205–209
mean cluster size 181
mean effective radii 146, 195 (footnote)

mean value theorem 187
medium size ring effect 198
molar concentration of rings, [G] 158, 161, 163, 167
molecular profile of an x-cluster 145
monodisperse melt 151–152
Monte Carlo simulation 183
m-tree 156, 158, 160, 170

net amount of rings 164
number fraction of x-clusters 145
numerical calculation 180

osmotic pressure, Π 147, 150

partition function, Z 147–149
percolation model 169
percolation threshold 191
perimeter polynomial 182
perturbation expansion 189
pivot algorithm 153
polycondensation 154
polyethylene oxide 201–202
probability of a FU being occupied by cyclic bonds, p_R 185–186, 191

radius of convergence 181, 183
random distribution assumption 186, 209
ratio abnormality 197
relative cyclization frequency, φ_j 158, 165, 195, 208
relative velocity 154, 169–170

reversible gel 189–190
ring closure probability, P_{cy} 156, 168–169, 171
ring distribution function 155, 158, 161
ring formation 154

scaled end-to-end distance, $|r|/\langle r_N^2 \rangle^{1/2}$ 153
screening effect 150
second virial coefficient 151
self-avoiding walks 153, 171
series expansion 181–183
site-bond diagram 205–208
surface area of a d-dimensional sphere, S_d 148, 156

thermodynamic equality 196–197
threshold 180–183, 190–193, 203
total number of chances for j-ring formation, ϕ_j 156–157, 160, 170–171
transition probability 154, 169
Tung-Dynes method 178–180

van der Waals formulation 146, 149
van't Hoff equation 147
Vinylone 141
Vinylone S 142
viscoelastic method 178
viscous liquid 178

weight average functionality 186, 191
weight fraction of rings, w 163

RETURN TO: **CHEMISTRY LIBRARY**
100 Hildebrand Hall • 510-642-3753

LOAN PERIOD	1	2	3
4			

2 HOUR

by your patron ID number.

DUE AS STAMPED BELOW.

MAY 1 8		

FORM NO. DD 10 UNIVERSITY OF CALIFORNIA, BERKELEY
2M 5-01 Berkeley, California 94720–6000